THE VOCABULARY OF SCIENCE

Some books by LANCELOT HOGBEN

Mathematics for the Million

Science for the Citizen

The Loom of Language (by Frederick Bodmer)
(edited and arranged by Lancelot Hogben)

Essential World English

Mathematics in the Making

The Mother Tongue

THE VOCABULARY OF

SCIENCE by Lancelot Hogben F.R.S.

With the assistance of

Maureen Cartwright

STEIN AND DAY / *Publishers* / New York

First published in the United States of America by
Stein and Day/*Publishers* 1970
Copyright © Lancelot Hogben 1969
Library of Congress Catalog Card No. 77-108314
All rights reserved
Printed in the United States of America
Stein and Day/*Publishers*/7 East 48 Street, New York, N.Y. 10017
SBN 8128-1287-5

235891

For F. A. E. Crew
to whom I owe so much

Acknowledgments

The author is greatly indebted to Professor E. H. Warmington for drawing attention to several errors and for making many helpful suggestions after seeing the proofs.

Dr R. H. Cawley gave much help in tracing sources mentioned in Chapter 2.

L.H.

Contents

PART ONE

1 The Latin Legacy

If we mean by science the written record of man's understanding of nature, its story begins five thousand years ago. Western science is thus a fabric to which threads of many colours have contributed before Britain, North America and Northern Europe were literate. Egypt and Mesopotamia, the Phoenician colonies and the Greek-speaking world of Mediterranean antiquity, the civilizations of China, India and the Moslem world supplied warp and woof in turn before Christendom began to make its own contribution.

Since the focus of expansion shifted to Western Europe less than five centuries before our time, the progress of the natural sciences has been spectacular in terms of both the immensity of now known facts about nature and the scaffolding of theories we invoke to interpret them. This is a commonplace; but few of us realize the uniqueness of one characteristic incidental to this efflorescence. Western Christendom has equipped what is now world-wide science with a world-wide and *constructed* vocabulary. It is no longer evocative to most of us. At expenditure of far less effort than the price exacted formerly, it could once more be so.

This world-wide vocabulary of Western science is the nearest thing to the lexicon of a truly global auxiliary that mankind has yet achieved. It derives its stock-in-trade almost exclusively from two dead languages, whence its word material is to most of us emotively neutral. Within the lifetime of the writer a smattering of both its components, i.e. Latin and Greek, was obligatory for entrance to many universities of the Western world. From the start, students of natural science thus held all the clues to decoding it. Most of them now have no knowledge of Latin, still less of Greek. To them, a vocabulary adequate to the needs of many branches of scientific enquiry is forbidding and mysterious.

It is not practicable to put back the clock to the days when study of Greek and Latin grammar as a prelude to translating Caesar or Ovid and the New Testament or Aristophanes was compulsory. There are now far too many curricular competitors with better claims. Even were it practicable, it would not necessarily be desirable in terms of

3

vocational training. The truth is that courses in classical studies formerly prescribed for students of natural science made demands on time and effort vastly in excess of their later needs. With the help of this book, every student of natural science should be able to gain in a few weeks more than a nodding acquaintance with the overwhelming majority of Latin and Greek words which occur as components of internationally current technical terms. Many names derive from Greek medicine or mediaeval Latin herbals and bestiaries. The author has arranged these in a separate chapter for the benefit of biologists and medical students.

That it is indeed possible for the reader of this book to gain insight into the rationale of scientific nomenclature with so little expenditure of time and effort is because he or she presumptively starts with the advantage of readiness *to make lively associations with already familiar terms*. Without recourse to the hints given in the glossaries of Chapters 5 and 6, many readers with no prior knowledge of Latin or Greek will recognize at sight as components of the following familiar words the Greek and Latin names of the four elements of antiquity:

(a) geology   terrestrial
     geography   subterranean
     geodesy   terrain

(b)   aeroplane
                aerial
                aeronaut

(c) pyrex   ignition
     pyrometer   igneous
     pyrotechnics   ignis fatuus

(d) hydrogen   aqueduct
     dehydrate   aqueous
     hydrometer   aquatic

In all countries where modern medicine, modern plumbing, modern agriculture and modern engineering penetrate, the vocabulary of science speedily makes its impress on daily speech. What was yesterday the jargon of the expert becomes an ingredient of the vernacular.[1] Thus scientific nomenclature is international in a dual sense. It is everywhere the vocabulary of the expert and its components are daily invading the speech habits of widely separated communities. In contradistinction to a process of word-building which can be in this way meaningful to people who speak different languages, military practice

[1] i.e. the home language.

and salesmanship have imposed a pattern which is exclusively national. It operates with initial letters of vernacular phrases as in *Operation PLUTO* (*Pipe Line Under The Ocean*) of World War II. The separate elements of this are not even suggestive to a person whose native language is English, and the interpretation is meaningless to a person unfamiliar with English. Of late, monstrosities of this sort have penetrated the laboratory. Such are *lasers* and *masers*.

The supreme merit of the traditional scientific recipe for creating new words for new things or new concepts comes sharply into focus, if we contrast the build-up of PLUTO (as interpreted above) with that of *telephotography*. Each of the three components of the latter occurs in many other words, e.g. *telescope, photon, graphite; telephone, photogenic, epigraph*. By comparison of different words in which each occurs, we are thus able to identify their meanings, and by permuting a very small number of such roots, we can generate a very large number of meaningful compounds. A simple calculation suffices to convey how immense a vocabulary one can generate by using a battery of only 200 such roots invoking: 1 (e.g. *photon*), 2 (e.g. *photograph*) and 3 (e.g. *micro-photography*) at a time. If we consider combinations only, there will be $^{200}C_1 = 200$ words of the first class, $^{200}C_2 = 19,900$ of the second and $^{200}C_3 = 1,313,400$ of the third, i.e. in all 1,333,500 – more than a million. If we allow for different permutations of the same roots (e.g. *phonogram* and *gramophone*), the corresponding total is:

$$200 + 39,800 + 7,880,400 = 7,920,400$$

Even allowing for the fact that some combinations or permutations would not be serviceable, it is evident that a basic vocabulary of 1,000 roots would suffice to generate very many millions of words.

Only so has it hitherto been possible *intelligibly* to keep pace with the need for new names resulting from the vast expansion of our knowledge of nature during the last two centuries. In A.D. 1450 the number of animals and plants with recognizably the same names for the same species throughout Western Europe was little, if at all, more than a thousand. In 1750, the number of plants with internationally current names was about 7,000, of named animals about 4,500 and of named substances less. By 1950 the number of named flowering plants and of beetles alone had each increased to about a quarter of a million, the number of named diseases to over a thousand, and the number of named organic compounds to more than three-quarters of a million including more than 10,000 dyes, of which over 1,200 were marketable.

Even if its meaning is not so transparent as are *geo-, photo-, phono-*

or *tele-*, the meaning of a Greek or Latin root contributory to the build-up of an internationally current technical term is easy to memorize by association therewith. Study of classical authors is therefore inessential to the best use of the world-wide vocabulary of science. Since it is constantly growing, it is, however, impossible to draw up a final list of what students or scientific workers may need to know without consulting a lexicon; and the proper use of a lexicon is not as simple as some may suppose. Though it is possible to use a Greek one intelligently without completing a four- or five-year sentence of grammatical punishment, it does not suffice to be familiar with the Greek alphabet and the customary conventions for transliterating from Greek to Latin (Chapter 4). A few facts about Greek and Latin grammar (Chapter 5) are also necessary.

It will help many of us to cherish as well as to make good use of it, if we take, in this chapter and the next, a backward look at our common linguistic heritage. Few of us know much about how we got where we are, and many of us are content to regard our birthright as the legacy of a defunct educational system. It is; but this is only half the story. The other half discloses a dilemma which led to a crisis the outcome of which the participants could not have foreseen. To understand it, we need to know how universities began in Western Europe and why Latin was their common language.

In A.D. 415, progress of pagan science came to a standstill in Christendom. It was the year of grace in which the sex-starved monks of St Cyril felled Hypatia from her lectern in Alexandria, stripped her, scraped her still quivering limbs with oyster shells and consigned what was left to the flames. Some four centuries later, the Caliph of Baghdad received from the Emperor of the residual Eastern rump of the Roman Empire, an ample collection of Greek MS. Syriac Christians, then tolerated throughout the Moslem world, translated the works of Euclid, Ptolemy, Galen and Aristotle into Arabic, and prepared the way for a new outburst of scientific knowledge to which the Hindu algorithms and Hindu trigonometry made a novel and lasting contribution. By that time, illiterate Teutonic tribes had overrun the western half of the Roman Empire, now cut off from African and Near Eastern coreligionists by the Moslem occupation of Spain, Sicily, Egypt and Palestine. Having imposed Latin as the medium of worship wherever it was in the ascendant, the Papacy had little sympathy with the language of its Byzantine competitor in the Christian stakes. In the West, the Norman Conquest of Britain, where it had flourished at the Celtic fringe, sealed the fate of Greek scholarship. Latin became the lingua franca of the monasteries of Western Christendom.

Within their walls, scientific knowledge had still two opportunities for a breakthrough. When Christianity became the imperial creed, its pastors had to assume the one useful social responsibility of their priestly predecessors. Like the priesthoods of pre-Christian antiquity, the new hierarchy became the custodians of the calendar. As such – and with all the authority of so doughty a detractor of pagan science as St Augustine – it had to find a niche for astronomy in its curriculum of permissible studies. At a later date, the monastic aspiration to benefit by the beatitude for those who tend the sick was a second breach in the walls. Monks founded hospitals and cultivated physic gardens. Where they did so, they were tolerant to Jewish missionaries of Moslem medicine. Jewish physicians practised at a time when Western Christendom as a whole was bitterly hostile to the Jewish stranger within the gates. They taught anatomy at a time when papal edicts prohibited dissection of the human cadaver.

Such was the situation in the century A.D. 1150–1250 when monks such as Adelard of Bath and Jewish scholars who had studied in the Moorish seats of higher learning in Spain, circulated Latin translations of the Arabic texts through which Western Christendom, like the Moslem world at an earlier date, had access to the teaching of Euclid, Ptolemy, Galen and Aristotle. These translations, especially (alas) of Aristotle, moulded the teaching curriculum of the European universities which took shape in the same period.

Before the *studium generale* – to give the fully fledged mediaeval university its contemporary name – took shape, a revival of learning in Italy was under way. Somewhat before A.D. 900, there came into being a medical school at Salerno. In close propinquity to Sicily, still occupied by Moslems before their final expulsion in A.D. 1091, Salerno was a springboard from which Moorish anatomy and pharmacy spread further afield – first to Montpellier about a century later. In Salerno, the teachers were mostly Jews, and some of them lectured in Hebrew. Thus Jewish physicians trained in Moorish medicine played a leading role as founders of the Montpellier school, located in the south of France with easy access by sea to Moorish Spain where Moslem science still flourished at Toledo, Cordova and Seville.

Independently, and before Western Europe harvested scientific knowledge of Alexandrian antiquity through Latin translations, an indigenous development occurred in Italy. About A.D. 1000, Bologna became a centre for the revival of legal studies based on the Roman code. In the two successive centuries, it attracted students in search of other sorts of instruction. By 1200, it had faculties of medicine and philosophy.

In the context of mediaeval Western Christendom, the term philosophy signifies the so-called *Seven Liberal Arts* distributed among two categories. The *Trivium* embraced *Grammar* (Latin), *Rhetoric* and *Logic*, the last two based on Aristotle's teaching. The *Quadrivium* embraced *Arithmetic, Geometry, Astronomy* and *Music*. Of these, the *Trivium* made straight the path for a *mariage de convenance* between Aristotle's teaching and dogmatic theology. Usually called *Scholasticism*, this retreat into obscurantism was the dominant *motif* of the university of Paris founded about A.D. 1160. Till A.D. 1360, Bologna had no faculty of theology recognized as such by the Vatican.

North and west of Italy and France, with or without papal charters, mediaeval universities came into being before A.D. 1400 – in Britain and at Prague before A.D. 1300, later in Cracow, Vienna and Heidelberg. Of the early universities of Western Europe, the birth of that at Lisbon in A.D. 1290 is noteworthy. It happened about a century after liberation of all we now call Portugal from Moslem occupation, and at a time when Moslem merchants enjoyed freedom to settle therein. Till the current dictatorship of Portugal deems it prudent to arrange publication of all the resources of its National Library, we shall not know how much its first university contributed to the most momentous contribution of Moslem culture to the enlightenment of Western Europe.

About A.D. 1420, Prince Henry of Portugal, known to posterity as the *Navigator*, set up a school of seamanship near Cape St Vincent. During the next forty years, he devoted himself to study of the considerable advances in scientific geography made by Moslem cartographers since the time of Ptolemy (*circa* A.D. 150). He enlisted Arab cartographers and Jewish astronomers to instruct his captains and to pilot the vessels which explored (a few years after his death) the coast of equatorial West Africa. There his successors built and fortified the castle of Elmina (A.D. 1482) now standing in Ghana. One of the captains of this expedition was Columbus. Undeterred by the Spanish Inquisition, Jewish pilots trained in Moorish nautical technology made a pivotal contribution to the rediscovery[1] of the New World.

When Henry of Portugal died in 1460, too early to receive news of how far Portuguese ships were soon to penetrate the Gulf of Guinea, the first presses had begun to print from movable metal type. Within a few decades, they were to produce nautical almanacs, commercial

[1] i.e. after the Icelandic expeditions of Leif Eriksson.

8

arithmetics expounding the new Hindu–Moslem algorithms and trigonometrical tables. Before the turn of the century the Columbian voyages had given a new impetus to the study of astronomy and scientific geography and to improved methods of computation demanded by both. In Italy, whose merchant princes had retained commercial and cultural links with the residual territory of the Eastern Empire, an influx of Byzantine refugees following the fall of Byzantium to the Turks quickened interest in the Greek language and examination of Greek texts which the printing press could now make available to scholars elsewhere.

Attracted by the reputation of a teacher in a particular discipline, European students of the sixteenth, as in the previous, century could migrate from one university to another. This was possible because Latin was everywhere the medium of instruction, and knowledge of Latin was the gateway to study. In mediaeval Europe, the Latin of the lecture room was colloquial; and at least as near to the daily speech of contemporary Italy as to the literary Latin of Tacitus. Encouragement of classical Latin as a status symbol in Renaissance Italy started a new fashion. It cannot have made life less easy for the itinerant pupil when teachers in Italy, among them Galileo, started to lecture in their own language. First of the societies formed to promote intercourse between inventors and investigators, the Italian *Lincei* set a fashion followed elsewhere. Like the *Lincei*, others conducted their proceedings in the vernacular.

When the great scientific awakening of the mid-seventeenth century enlisted in a common endeavour master pilots, watch-makers, spectacle makers, master gardeners such as Thomas Fairchild of Hoxton and tradesmen such as Leeuwenhoek of Delft, the Reformation had displaced Latin as the medium of worship in countries where ecclesiastical obstruction to scientific progress was minimal. By no means all the newly recruited personnel devoted to promotion of naturalistic knowledge were now steeped in Latin scholarship. In this milieu, the British Royal Society and the French Academy followed the lead of the *Lincei* by adopting the vernacular alike for oral and for written communication. In 1687, Newton had published his *Principia* in Latin. Seventeen years later, his *Opticks* appeared in English.

Albeit inevitable and beneficial in its own setting, the decision to do so carried with it a penalty. Hitherto physicians, chemists, astronomers and mathematicians of Western Europe had used Latin as the medium of scientific publication. Henceforth there was no lingua franca in which men of science of different speech communities could

9

communicate their discoveries. To the lively sense of companionship in a common enterprise between men such as Hooke and Huyghens, Conmenius, Torricelli, Mersenne, Gassendi and Leibnitz, severally in Britain, Holland, Bohemia, Italy, France and Germany, the lack of a single medium of communication was intolerable. Leaders of science were thus alert to the need for a common language, the more so for two reasons. There was now urgent contemporary concern for standardization of algebraic symbolism, and Jesuit missionaries had lately brought to the notice of European scholars the ideographic script in which scholars with no common speech could communicate by visual symbols.

In Britain, the Royal Society (1664) officially commissioned Wilkins to seek a remedy. A liberal bishop, who had been chairman at its inaugural meeting, Wilkins brought to the task expert knowledge of cryptograms and familiarity with a shorthand lately introduced to record state trials. On the Continent, Leibnitz approached the issue in terms of his search for a universal algebra of reasoning. He devoted much of the leisure of his later years to the same end; but he did not live to complete his project. Before the death of Leibniz, the *Real Character* of Bishop Wilkins was already in print. Published in 1668, it was not actually the first constructed auxiliary. Seven years earlier, George Dalgarno of Aberdeen, author of a deaf and dumb sign language, had published his *Ars Signorum*, a system remarkably like the *Real Character*, each with Chinese overtones. Wilkins himself denied prior knowledge of it. It is charitable to surmise that both authors were familiar with a programme outlined by Descartes and anticipating features common to both projects.

Descartes had hoped that human ingenuity could construct a language fit for peasant and philosopher alike; and he recognized that an attempt to reinstate Latin could satisfy the requirements of neither. Though Dalgarno and Wilkins had a less ambitious aim, neither the *Ars Signorum* nor the *Real Character* gained widespread support. They had far less mnemotechnic merits than the Latin they were to supersede, and a quasilogical structure common to both provided no room for further expansion of knowledge. For two centuries thereafter, interest in the promotion of an auxiliary medium was dormant, while men of science in the following of Linnaeus and Lavoisier contented themselves with a more limited objective. They set about piece-meal reform of the vocabulary of science itself.

Reforms initiated by Linnaeus and Lavoisier will be the topic of the next chapter. To assess rightly the contribution of the former, we need to bear in mind that Latin remained the written language of science in Scandinavia and Germany long after the vernacular had displaced it in Italy, Britain and France. Well into the first half of the nineteenth century, Gauss continued to publish the results of his mathematical researches in Latin, and classical Latin at that. Their importance might have gained earlier recognition had he published them in German.

At the end of the eighteenth century, most of those who used elsewhere in Europe, as in the American colonies, the vernacular as a medium of scientific communication had still at least a smattering of schoolboy Latin. Even those who had not, inherited a vocabulary from the time when Latin had been the lingua franca of scholarship throughout Western Christendom, and by doing so they had unwittingly assimilated a vocabulary which was not exclusively Latin. Before Linnaeus and Lavoisier, men of science had little disposition to exploit the resources of the Greek language to coin new terms. By using Latin during a period when there was no Greek scholarship in Western Christendom, they had however assimilated, along with its store of scientific knowledge, the technical jargon of Greek-speaking antiquity.

Roman civilization being what it was, this was inevitable. Though glamorized by the legally minded as the *fons et origo* of the Rule of Law, Roman civilization had the defects of a slave-owning society. Persistence of the gladiatorial show was incompatible with the modicum of compassion favourable to medical care; and abundant cheap labour stifled incentive to mechanical ingenuity.[1] The Roman rule of law was a sideline to a career of aggression of duration with no counterpart in the social history of mankind. As part of the pay-off, it replenished a labour force of slaves for the mines or for construction of their aqueducts and treated it with harshness rarely exceeded by earlier civilizations.

Apart from the maintenance of a priestly calendar in frequent need of revision by recourse to the practice of less backward communities, what little science the Romans assimilated from their betters was useful only as a means of promoting more military exploits with brighter prospects of enslavement. Even so, the Latins were at a loss for words. It is on record that the meagre resources of so technologically primitive a community could not provide a vocabulary adequate to convey the fertile speculations of the Ionic Greeks or to preserve the positive

[1] Archimedes, who was butchered by Roman soldiers established the basic principles of hydrostatics about a century before the praetor Marcius ordered the construction of an aqueduct nearly sixty miles long. By the end of the first century A.D., nine such horizontal watercourses supported by gigantic stone pillars supplied Rome with water at vastly greater expense than that of laying down clay pipes for a liquid to find its own level.

achievements of Alexandrian science. Listen[1] to the words which Lucretius, author of the first exercise in popularization of science, addressed a reluctant audience of Roman slave-owners:

> I know how hard it is in Latin verse
> To tell the dark discoveries of the Greeks,
> Chiefly because our pauper speech must find
> Strange terms to fit the strangeness of the thing.

The Romans borrowed Greek words to transmit what Greek science they did preserve. By the time the Western Empire broke up, they had indeed assimilated a considerable Greek vocabulary. Because their alphabet was not identical with that of the Greeks, they imposed on Greek words their own spelling conventions. Partly because Greek pronunciation changed during six centuries of Roman rule and partly because latter-day authors supplemented the Roman alphabet with Greek letters, such conventions were not wholly consistent when handed down to posterity. None the less, a Greek component of the world-wide vocabulary of Western Europe was already latent in the Latin used by men of science in mediaeval Christendom. This is one reason why Western science turned to Greek when it had run short of verna-cular vocabulary resources in the eighteenth century of our era.

The reader may here ask why Latin remained the medium of scientific communication in Teutonic countries so long after men of science in Italy, Britain and France had come to terms with everyday speech. A plausible, if incomplete, answer to this question throws light on Anglo-French collaboration on the reform of scientific nomenclature during the closing years of the eighteenth century.

It goes without saying that transition from colloquial Latin to the vernacular of Galileo's Italy called for no drastic break with traditional habits. Though French terminals are less conspicuously than Italian like those of the common parent, the adaptation of a Latin noun or adjective to the French form could call for little mental effort and encountered no obstacle from an alien flexional system such as that of German. Thanks to the Norman Conquest and nearly four subsequent centuries of dynastic wars in which Englishmen fought on French soil, the English language of Tudor times had already appropriated a well-nigh exhaustive battery of Latin terminals more or less modified by French usage. Even when the *Mayflower* sailed, the language of the Pilgrim Fathers had assimilated through French (e.g. *royal, loyal*) or

---

[1] *De Rerum Natura*, as translated by W. Ellery Leonard (1916). The date of the original is uncertain – between 50 and 99 B.C.

directly from classical authors (e.g. *regal, legal*) an enormous equipment of Latin descriptive terms, and there was a *well-established pattern for adapting any newcomer from the same source.*

To interpret rightly the revolution of scientific nomenclature due to the work of men such as Linnaeus and Lavoisier, we shall need to invoke this pattern. So we may usefully pause to glance at some of the suffixes which make it possible to adapt the Latin vocabulary of mediaeval scholarship to English usage:

-ACEOUS	Latin -*acea* (nom. fem. sing.); French -*ace* (masc. sing.) as in: *foliaceous, herbaceous*
-AL	Latin -*ale* (ablat. sing.); French -*al* (masc. sing.) as in: *lateral, radical*
-ANT *and* -ENT	Latin -*ante* and -*ente* (abl. sing.); French -*ant* or -*ent* (masc. sing.) as in: *rampant, sentient*
-AR *and* -ARY	Latin -*aria* (nom. fem. sing.); French -*aire* (masc. sing.) as in: *lunar, regular* and *arbitrary, rotary*
-ATE	Latin -*atea* (nom. fem. sing.) as in: *ornate, sedate*
-ATION *and* -ITION	Latin -*atione* and -*itione* (abl. sing.); French -*ation* and -*ition* as in: *nation, ignition*
-FEROUS	Latin -*fer* (nom. masc. sing.); French -*fère* (masc. sing.) as in: *carboniferous, coniferous*
-FIC	Latin -*fica* (nom. fem. sing.); French -*fique* (masc. sing.) as in: *soporific, terrific*
-FID	from perfect tense of Latin *findere* (to split) as in: *bifid, pennatifid*
-FORM	Latin *forma* (nom. fem. sing.); French *forme* (shape) as in: *uniform, vermiform*
-IC	Latin from Greek – *ικη* (nom. fem. sing.) as in: *historic, emetic*
-ICAL	= -IC + -AL (*above*) as in: *historical, umbilical*
-ILE	Latin -*ile* (abl. sing.); French -*ile* (comm. sing.) as in: *servile, sessile*
-INE	Latin -*ina* (nom. fem. sing.); French -*ine* (fem. sing.) as in: *canine, pristine*
-ISION	Latin -*isione* (abl. sing.) as in Fr. and Eng: *collision, erosion*
-ITE	Latin -*ita* from Greek – *ιτη* (nom. fem. sing.); French -*ite* (fem. sing.) as in: *erudite, tripartite*
-OID	Latin -*oide* (abl. sing.) from Greek – *οειδη* as in: *rhomboid, ovoid*
-OUS *and* -OSE	Latin -*osa* (nom. fem. sing.); French -*euse* and -*ose* (fem. sing.) as in: *gracious, ligneous* and *grandiose, varicose*

Against the background of the hybrid heritage of the English speech community, we can readily grasp the reluctance of German scientists to abandon the use of Latin. To the extent that one can meaningfully speak of any language as such, German of the eighteenth and nineteenth century was a *pure* one. Daily speech had few traces of borrowed Latin; and the highly flexional structure of the written word was a formidable barrier to intrusion of alien words. Partly at least, this accounts for a chasm of incommunicability between the German professor and the general public at a time when such men as Faraday could attract enthusiastic audiences of London artisans. Contrariwise, the hybrid nature of their mother tongue fostered communication between British men of science and a public eager for information. From the days of Davy to the present time, Britain has had a long tradition of popularization in which some of her most eminent scientific workers have actively participated.

What is more relevant to the theme of this book is another consequence of our hybrid heritage. In the last few decades of the eighteenth century British pioneers of chemical industry shared a common vocabulary with French chemists in the forefront of language reform. This is peculiarly true for a reason which the foregoing table of terminals helps us to understand. Even before the eighteenth century, English had assimilated such words as *arsenic* (Middle English) and *cupreous* (1666); but personal taste alone dictated whether to tack on *-ous* or *-ic* to a root. A king-pin of the Lavoisier reform was to endow such suffixes with a meaning relevant to *quantitative recipes* for making new substances. When the French pioneers of language planning made such rules, e.g. to distinguish between nit*ric* and nit*rous* acids, or between sulph*ates* and sulph*ites*, they could enlist in a common endeavour investigators of the two nations then in the vanguard of chemical discovery. This was possible because, and only because, with minor differences of spelling (E *-ous* for F *-euse*, E *-ic* for F *-ique*), the participants could draw on a common stock of terminals. Had Britain been at that time in the rearguard and Germany in the forefront, the prospects for international acceptance of the French programme would have been bleak (see pp. 34–36).

2 The Reinstatement of Greek

During the first of two centuries (1650–1850) which witnessed the rise of Britain and France to world leadership of scientific discovery, biological taxonomy was the pacemaker for adapting to a common vernacular (p.4) pattern an expanding vocabulary of technical terms bequeathed by the Latin of mediaeval scholarship. Before the middle of the seventeenth century the nomenclature of systematic botanists and zoologists had assimilated through Latin a substantial glossary of Greek roots at second hand. Thenceforth, it gave a powerful impetus to the use of others for names of new things and new concepts. Without a nodding acquaintance with the history of botanical and zoological classification, it is thus difficult to appreciate the *raison d'être* of one aspect of the reform of chemical nomenclature in the two closing decades of the eighteenth century.

Since our concern in this chapter will therefore be with how Western science renewed its debt to Greek civilization, it is fitting first to take a backward look at what Western civilization owes to Greek science. When one speaks either of the debt of Western civilization to Greek science or of the debt of modern science to Greek civilization, one is not speaking of what we owe to a nation, still less to a nation mainly located on the European mainland west of the Bosporus and the Dardanelles. What one customarily calls Ancient Greece was an assortment of city states, monarchies and mercantile colonies extending from the coastal region of Asia Minor to the western margin of the Mediterranean. Besides the mainland of modern Greece, it included a large part of the toe of Italy, with Sicily, Crete, Cyprus and a multitude of smaller islands and coastal settlements as far afield as Marseilles. Greek-speaking communities, often at war among themselves and with no lasting framework of alliance in times of peace, had an overall government only under the brief rule of Philip of Macedon and that of his son Alexander, whose generals established dynasties in Egypt, Mesopotamia and Syria after his death.

The cosmopolitan city of Alexandria, built in 332 B.C. to celebrate Alexander's conquest of Egypt, was in close propinquity to one of the two great temple repositories acknowledged in antiquity by the island

Greeks for its mathematical lore and astronomical science. For two centuries it had cultural links with the Greek-speaking descendants of Alexander's veterans in Mesopotamia, where the sister sciences of the temple precincts had attained a higher level than in Egypt. In what we may rightly call the university of Alexandria, the Greek language became the vehicle of an astonishing efflorescence of scientific discovery enduring from the installation of its first notable teacher Euclid (*circa* 300 B.C.) to the death of Theon (*circa* A.D. 400), last of its mathematicians and father of Hypatia (p. 6). These seven centuries circumscribe the major part of the Greek contribution to modern science salvaged by Moslem scholars whose translations also transmitted to the mediaeval universities of Western Europe what Alexandrian men of science testified to the contribution of their maritime predecessors.

In short, the only feature common to all we mean when we speak of Greek science is scientific knowledge initially transmitted through the medium of the Greek language; and the word *maritime* in the preceding sentence puts the spotlight on where it flourished most conspicuously before the Alexandrian episode. The lasting contribution of Greek science to posterity had indeed little to do with what the poet calls the Glory that was Greece, meaning thereby the period when Athens was the birthplace of a memorable dramatic literature and a Mecca for millionaire playboys attracted by the glamour of debating contests in philosophical speculation. For the future of science, the Athenian drama had at least as much to contribute as the speculative exercises of the Academy and the Lyceum. Its dialogue brought the written word closer to everyday speech than ever before and equipped scientific discovery with a novel and vastly liberating tool of communication.

Though the importance of the new instrument of communication is difficult to overstate, posterity has been prone to exaggerate the positive contributions of the schools[1] of Plato (*circa* 390 B.C.) and Aristotle (*circa* 350 B.C.). Admittedly, the colleagues and pupils of Plato consolidated the logical foundations of geometry, but they did so with a compass and rule *ukase* which stifled curiosity and discouraged the art

---

[1] The Athenian school of Epicurus (*circa* 300 B.C.) is noteworthy because it adopted the atomistic views of Democritus and, through the biography of Diogenes Laertius, is our main source of information about them. However, the main concern of Epicurus was to formulate a quasi-rational theory of morals. His school initiated no programme of experimental enquiry to explore new ways of testing the particulate theory of matter. Unlike that of Democritus, the outlook of Epicurus himself was merely argumentative in the Socratic tradition of Plato.

of measurement. The attitude of the maestro himself to naturalistic enquiry was haughtily hostile. Unlike Plato, Aristotle encouraged naturalistic studies; but his treatment of astronomy and scientific geography added little, if anything, to what maritime Greeks of earlier[1] vintage had discovered.

To be sure, Aristotle's treatise on the comparative anatomy of animals, known best by its Latin title *Historia Animalium*, may deserve personal credit for extensive and careful observations. Even so, the impact of his teaching on that of the mediaeval universities of Western Christendom was largely retrograde. He dismissed valid and incontrovertible evidence which led Empedocles (*circa* 475 B.C.) to conclude that air has weight. Therewith, he rejected the cogent case advanced by Democritus (*circa* 430 B.C.) in support of a particulate theory embracing air and vapours as a third – what we now call the gaseous – state of matter. He also bequeathed to posterity a *mystique* which discouraged a rational approach to both the study of terrestrial gravitation and an understanding of combustion.

During the period when Catholic Europe, through the spread of Latin texts based on Arabic translation, was assimilating the positive contributions of Greek science along with Aristotle's logic and cosmogony, Greek studies had languished under papal discouragement and the mediaeval universities offered no instruction in the Greek language. Meanwhile, Italy retained lively mercantile intercourse with the shrinking Byzantine rump of the Roman Empire. This set the stage for the cultural upheaval known as the Italian Renaissance and as Humanism elsewhere in Europe. From the beginning of the fifteenth century onwards, the Italian plutocracy encouraged closer cultural ties with Constantinople, and Byzantine refugees had brought with them into Italy a substantial stock of Greek manuscripts during the half century before the beleagured city finally capitulated to the victorious Turks (1453 A.D.). Eleven years later, printing from movable type began in Italy, and Italian master printers made Greek texts available to scholars throughout Europe.

A few of the latter, like Linacre of Oxford, were men of scientific bent; but the spread of Greek scholarship had little immediate effect on scientific nomenclature. Its main impact was to intensify the religious

[1] Greatest of all the geographers, before Eratosthenes of Alexandria (*circa* 250 B.C.) measured the circumference of the earth with an astonishingly small error, was Pytheas, a master mariner of Marseilles. Being twenty-four years younger than Aristotle, he was young enough to have studied under him when Aristotle was teaching in Athens. It is almost certain that they never met, still less that he was a pupil of the Stagyrjte.

ferment of the time. During the sixteenth century, new translations of the Bible and of patristic literature based on Greek texts become powerful weapons of theological controversy in opposition to papal authority; but no Greek writer unknown to Moslem scholars exerted a profound influence on scientific thought before commentaries of Gassendi on Epicurus (1647–9) introduced men of science to the atomic speculations of the early Greek materialists. This did not happen till well over a century after the first German (Luther, 1522) and the first English (Tyndale, 1526) translation of the Greek New Testament had appeared in print.

When Gassendi introduced the generation of Hooke and Newton to the particulate theory of Democritus, the English language had absorbed directly from Greek few scientific terms. The following are examples introduced before 1750 with dates as cited by the *Shorter Oxford Dictionary*:

therapeutic	(1541)	nephritis	(1580)
hydrophobia	(1547)	physics	(1589)
genus	(1551)	phthiriasis	(1598)
theorem	(1551)	rhododendron	(1601)
rhythm	(1557)	magnetism	(1616)
physiology	(1564)	thermometer	(1633)
tetrahedron	(1570)	telescope	(1648)
pentagon	(1570)	microscope	(1656)
polygon	(1570)	barometer	(1665)
hexagon	(1570)	hyperbola	(1668)
theodolite[1]	(1571)	hydrometer	(1675)
phenomenon	(1576)	microphone	(1683)
chrysanthemum	(1578)	psoriasis	(1684)
parabola	(1579)	electrometer	(1749)

In 1750, assimilation of Greek roots by scientific nomenclature was proceeding apace. What had been a trickle in 1700 had become an avalanche in 1800. Part of the explanation is the colossal number of newly discovered objects and processes for which it was necessary to provide names. These included living creatures and their parts, synthetic substances, units of measurement, instruments and other inventions. The first of these categories is of special interest. Biological classification was largely responsible for the new fashion; but circum-

---

[1] The origin of this word is obscure. The *Concise Oxford Dictionary* suggests that it is a misspelling. The word θεα means a *spectacle, sight* or *view*. The word δηλος (*fem.* δηλη), as in *psychedelic*, means *manifest* or *clear*. An intelligible rendering would be *theadelite*. It can have no connexion with θεος (*god*) as in *atheist* or *theology*.

stances which prompted botanists and zoologists to quarry a new mine for word-making did not become clamorous till over a century after Greek scholarship had found a foothold in European centres of higher learning.

Before the end of the sixteenth century of the Christian era, biological science had advanced little since the time when Herophilus of Alexandria (*circa* 300 B.C.) expounded anatomy by dissection of the human body. The *De Fabrica Humani Corporis* of Vesalius (A.D. 1543) corrected some errors in Galen's teaching (*circa* A.D. 180), and set a new standard of realistic illustration due to the anatomical zeal of Renaissance painters: but knowledge of comparative anatomy at that date did not appreciably exceed the scope of Aristotle's *Historia Animalium*. Belief in their allegedly, and mostly spurious, medicinal value prompted the preparation of the first recorded catalogues of plants, that of Theophrastus (372–287 B.C.), the successor of Aristotle as teacher of his peripatetic school in Athens, and that of Dioscorides, a physician who served in the army of Nero during the latter half of the first century A.D. The *Materia Medica* of the latter incorporates and adds little to a *History of Plants* by the former. It describes in the vernacular about 600 species.

In Elizabethan England, Gerard's *Herbal* (1597) listed only 1,035 but a little earlier (1583) his contemporary Caesalpinus had described 1,520 plant species. A century later, J. P. de Tournefort listed about 8,000. The number of animal species mentioned by Aristotle is about 600, and the number of named animal species in A.D. 1600 did not greatly exceed 1,000, a quarter of the figure known to Linnaeus in the mid-eighteenth century.

It is thus true to say that any advance of biological knowledge between 300 B.C. and A.D. 1600 was trivial compared with progress between A.D. 1600 and 1750. Three circumstances conspired to inflate the vocabulary of the biological sciences during this century and a half:

(i) commercial horticulture received a powerful impetus both from the Dutch tulip industry, already flourishing in the sixteenth century, and from the spread of root-crop production;

(ii) colonization of the New World and expansion of trade with the Far East brought to the notice of physicians and gardeners plants hitherto unknown to them, some of actual (*cinchona* bark) or supposedly medicinal value, some which had ornamental appeal (e.g. *Pelargonium*) and others which were edible (e.g. potato, Jerusalem artichoke);

(iii) the lately invented microscope delved into a new domain of living creatures, disclosed new horizons of anatomical study and shed a flood of light on the reproductive process of animals at a time when men of science firmly believed in the goose–barnacle legend of Gerard (*see below*).

Apart from (iii), the foregoing circumstances are especially relevant only to the progress of botany. Before A.D. 1600, the *herbal* had been primarily a pharmacopoeia and the impulse to catalogue plants had been almost exclusively concerned with their medicinal use. Though the New World whaling and fur trade prompted explicit instructions of Elizabethan seamen to record information about possibly useful animal species, one cannot discern any powerful impetus other than idle curiosity to encourage publication of mediaeval catalogues (*bestiaries*) of animals. In Renaissance Italy, a private menagerie was the status symbol of a merchant prince. This may have conferred ostentation value on Italian patrons who paid for compilation of descriptions of foreign animals often from hearsay and sometimes as imaginary as the Loch Ness Monster.

The goose–barnacle legend is characteristic of what naturalists regarded as credible before the use of the microscope disposed of Aristotle's teaching, i.e. that the production of some animals 'is spontaneous. For some of them sprung from the dew which falls from plants. Some originate in rotten mud and dung. . . . Gnats originate in threadworms and the threadworms originate in the mud of wells. . . .' Gerard's account of the legend is as follows:

> There is a small island in Lancashire wherein are found the broken pieces of old and bruised ships . . . and also the trunks and bodies with the branches of old and rotten trees cast up there likewise, whereon is found a certain spume or froth that in time breedith unto certain shells in shape like those of the Muskle, wherein is contained a thing in form like lace of silk . . . one end of which is fastened unto the inside of the shell even as fish of Oisters and Muskels are: the other end is made fast unto the belly of a rude masse or lumpe which in time cometh to the shape of a bird; when it is perfectly formed the shell gapeth open and the first thing that appeareth is the aforesaid lace or string; next come the legs of the bird hanging out and as it groweth greater it openeth the shell by degrees, til at length it is all come forth and hangeth onely by the bill: in short space after it cometh to full maturitie and falleth into the sea where it gathereth feathers and groweth to a fowle bigger than a Mallard and less than a goose having blackle legs and bill or beak and feathers blacke and white spotted in such manner as is our magpie. . . . We conclude and end our present Volume with this Wonder of England. For the which God's name be ever honoured and praised.

For one or other reason mentioned above, compilations of plant or animal catalogues in the seventeenth century faced an increasingly formidable task of naming new species, of describing them intelligibly and of arranging them for identification with the minimum of time and effort. Some individuals engaged in the task travelled far afield, enlisting correspondents to assist them. Others made intensive studies of the flora and fauna of their own countries or districts. Classification became a sheer necessity of accumulating vastly more information than the human memory can accommodate. Meanwhile, difficulty of finding names for species having as yet none was scarcely less onerous than providing definitive labels for species which had the dubious advantage of already having vernacular names. If the range of a species extended over different speech communities, names intelligible in one would not be intelligible in another. Even within one speech community, folk names for the same species may be different in different localities. Thus Scots *blaeberries* are *wimberries* in Welsh border counties, *bilberries* in some parts of England and *whortleberries* in others.

A first step towards unequivocal designation was also a first step in classification. Before Linnaeus (1735) published the first edition of his *Systema Naturae*, many botanists and zoologists adopted at times, but not consistently, a practice which he regularized. This gave to every organism two names: a first (*generic*) name to indicate its place in a niche (*genus*) reserved for other organisms with closely similar characteristics, and a second (*specific*) name to identify it as one of a self-perpetuating assemblage (*species*). Thus Ray's *Synopsis Methodica* (1693–1713), which sets forth in two volumes a comprehensive classification of animals, distinguishes the lion, tiger and domestic cat respectively as *Felis leo*, *Felis tigris* and *Felis cattus*.

Before 1600, there had been minor exercises in the arrangements of plant and animal species for convenience of identification, notably the *Catalogus Plantarum* (1542) and *Historia Animalium* (1551–8) of Conrad Gesner, the Herbal of Jerom Brock (Hieronymus Tragus) published in 1551 and the *De Plantis libri XVI* (1583) of Caesalpinus. A herbal of the German botanist Joachim Jung published (1662) five years after his death carries the process of grouping species for ready reference a considerable step further. He arranged them in named assemblages such as *Compositae* (daisy family), *Labiatae* (dead-nettle family) and *Leguminosae* (pea family). Later authors found that naming all the species assigned to even one such group was a sufficiently exacting task. Thus Morrison (1672) published a treatise devoted only to the *Umbelliferae* – a group which includes several edible forms such

21

as the carrot, parsnip, carraway and angelica, several poisonous ones such as the hemlock of Socrates and, needless to say, several of alleged medicinal value.

Ten years later, John Ray published a memoir (*Methodus Plantarum*) on plant classification. He extensively revised it in 1703 with the title *Methodus Plantarum emendata et aucta*. Much water had flowed under the bridges since it first appeared. In particular, the recognition of sexuality in flowering plants had received experimental confirmation leading to the production of the first artificial plant hybrid, a Pink by Sweet William cross called Fairchild's Carnation after its producer. Ray's memoir of 1703 recognizes, and designates as such, the division of flowering plants into *Monocotyledons* and *Dicotyledons*, and it groups many genera into divisions corresponding to what we now call by the same names (e.g. *Cruciferae, Umbelliferae*) as *natural orders*.

From Aristotle and Greek medicine, Ray's generation had inherited a substantial vocabulary of names for animals and for their organs. Ray himself designates some of his divisions of the animal kingdom by phrases (e.g. *ventriculis praeditos duobus* = mammals and birds). For others he uses single names, of which *Ungulata, Aves, Pisces, Insecta, Crustacea* and *Malacostraca* survive, the last alone being of Greek origin. Other divisions, with names of Greek ancestry, but no longer in use, are those of the Ungulates as *Monochela, Dichela* and *Tetrachela*.

The first volume of the *Systema Naturae* of Linnaeus published in 1737 has a more modern aspect than the *Synopsis Methodica* of Ray; and the work as a whole has a far more ambitious aim. Its author divided the *Imperium Naturae* into three kingdoms: animal (vol. 1), vegetable (vol. 2) and mineral (vol. 3). In the first of these, he retained some Latin names for groups and introduced others, a few of which (e.g. *Mammalia*) have survived. He also introduced a considerable number of group-names based on Greek roots. These include: *Amphibia* and *Zoophyta* together with names for orders of insects, namely, *Coleoptera, Hemiptera, Lepidoptera, Neuroptera, Hymenoptera, Diptera*. All of these are still in use. The arrangement of the *Regnum Animale* was more consistently hierarchical than that of Ray. Linnaeus assembled species in a threefold tier, first in genera, genera in orders and orders in six main classes: *Mammalia, Aves, Amphibia, Pisces, Insecta* and *Vermes*. He made no use of divisions (*phyla*) larger than classes nor of smaller ones (*families*) within orders introduced by his successors from Cuvier onwards.

In terms of its impact on the vocabulary of science, the second volume of the *Systema* is of greater importance than the first. That its

22

singular merit has gained scant recognition may be due to a barren controversy concerning the advantages of two ways of approaching the task of classification. To this day, some naturalists continue to convey the misconception that one can meaningfully speak of taxonomical systems as right or wrong in contradistinction to being more or less relevant to the end in view. The detractors of Linnaeus thus blame him irrelevantly for renouncing the quest for a correct (*natural*) to create a false (*artificial*) system.

By *artificial*, one here means a serviceable key which proceeds step by step in a process of exclusion. This implies that each division is a class in Aristotle's sense, i.e. an assemblage of organisms which share one or more characteristics absent in all others. Such indeed is how one has to proceed in the final stages of identifying a species. However, one can make a short cut to the same goal if one recognizes so-called *natural* groups (e.g. *Rosaceae* or *Crustacea*) of which the members may hang together by transitional types without having any single characteristic uniquely peculiar to all of them. The immediate predecessors of Linnaeus were groping towards a system which recognized such assemblages; but the hierarchical arrangement of the divisions of plants in the *Systema Naturae* proceeded consistently along the path prescribed by Aristotle's logic. The process of exclusion took no account of groups with a recognizable family resemblance, albeit a family likeness not necessarily expressible by all-or-none criteria.

In short, the classification of plants by Linnaeus steered an even course of rejection or retention by grouping plants in categories of which constituent species with no other family resemblance shared only one unique character. Being in this sense artificial, his system was therefore defective as an economical way of identifying a plant by first excluding many species with a recognizable family resemblance for which Aristotle's two-way traffic furnishes no formula. Actually, few would seriously argue that it is practicable to eliminate all *artificiality* from a serviceable system. If one did so, one would cease to speak of *algae* and *fungi*, since the one common characteristic (presence or absence of assimilatory pigment) is of trivial[1] interest compared with the many characteristics some fungi uniquely share with some algae.

This artificiality of taxonomical procedure should not blind one to a conspicuous merit of the second volume of the *Systema*. The

[1] Most taxonomists place *Cuscuta* (dodder) in the *Convolvulaceae*. If we took the distinction between Algae and Fungi seriously we should place Cuscuta (with Orobanche, etc.) in a class apart from all other flowering plants.

predecessors of Joachim Jung and Ray inherited from the pharmacopoeias of Greek medicine and of their Moorish successors no corpus of plant anatomy and no standardized nomenclature of other descriptive terms. Jung recognized the need, but did not live to participate in the recognition of the sexual process and of its relevance to the task of equipping plant taxonomy with such a descriptive nomenclature. The supreme merit of the Swedish savant was that he undertook a meticulous classification of descriptive terms to clarify his classificatory system for plants. At the beginning of the second volume of the *Systema Naturae*, first published in 1735, we encounter a complete glossary of all adjectives used: (*a*) to label characteristics of different parts of a plant; (*b*) to describe its habit, site and uses, if any.

The former appear separately under the headings: RADIX (*root*); TRUNCUS (*shoot*); FOLIA (*leaves*); FULCRA (*bracts, petioles,* etc.); FRUCTICATES (*floral organs*). Altogether the list contains about 950 terms. Though Linnaeus wrote in Latin, English, like French (see p. 13), could provide appropriate suffixes for almost every one. The following item shows how readily English and French readers can thus adapt to the vernacular the specification of any species without prior knowledge of the Latin language.

676 DIGITALIS	*Cal.* 5-partitus. *Cor.* campanulata, 5-fida ventricosa. *Caps.* ovata, 2-locularis.
purpurea 1. D.	calycinis foliolis lanceolatis, corollis obtusis: labio superiore integro.
lutea 2. D.	calycinis foliolis lanceolatis, corollis acutis: labio superiore bifido.

In English this would read

676 DIGITALIS	*Calyx* 5-partite. *Corolla* campanulate, 5-fid swollen. *Capsule* ovate, bilocular.
1. *D. purpurea*	calyx leaves (*sepals*) lanceolate, corolla *ditto* (*petals*) obtuse, superior labium undivided
2. *D. lutea*	sepals lanceolate, petals acute, superior labium bifid.

Another unique feature of the second volume reinforced the new trend in scientific nomenclature, that of turning directly to Greek for word-building. Whereas the descriptive vocabulary is exclusively Latin, Linnaeus based the names of all the major divisions of the classification system on Greek roots. In thus starting a new fashion, he put all his cards face uppermost on the table. For the benefit of the reader who was ignorant of Greek, he cited under each class specification

Greek words from which he derived the roots incorporated in the class name, as illustrated by the following entry against the name of the fourteenth of his twenty-four major divisions:

XIV DIDYNAMIA a δις *bis* & δυναμις *potentia*

To Linnaeus, we must therefore concede credit for being the first man of science:

(a) to dispense deliberately with what Lucretius called the *pauper* speech as a source for enlarging the vocabulary of science;

(b) to prescribe a pattern for assimilating Greek roots to terminals (e.g. -IA of the above) with a unique taxonomical status e.g. (by his successors) -IDAE for animal families and -ACEAE for plant orders.

In both respects he anticipated the great French reform we shall come to later. That he can also claim in some measure to have inspired it, is partly due to the utter irrelevance of the third volume to a useful classification of inert matter.

With but one, albeit belated, bequest of lasting value to science, the classification of the mineral kingdom of the *Imperium* was a gigantic waste of effort. It classified the shape of such crystalline substances as occur in nature (*diamond, emerald, quartz*) in terms of Greek geometry; and mineralogy, still taught in 1914 as a discipline in its own right at Cambridge and elsewhere, cherished this crumb of useful taxonomical industry. Its pay-off came when the Braggs and their collaborators were able to exploit X-ray spectrography. Linnaeus held professional posts in Uppsala first as a physician, then as a botanist, and his passion for classification extended to diseases. His *Genera Morborum* published after the *Systema* was the parent of many other largely useless systems of diagnosis in the bedside manner.

As a professor of medicine at a time when the pharmacist was the precursor of large-scale chemical industries emerging from 1746 onwards, Linnaeus might more usefully have turned his attention to the theme of our next chapter. If the title of the *Systema Naturae* excludes the treatment of man-made products, and therefore any immediate contribution to the science of chemistry, the grand design of the author's *Imperium* does not account for so conspicuous an omission. Chemistry was indeed his blind spot. At the date when the *Regnum Minerale* first appeared, chemists already recognized several different sorts of air – or, as we now say, *gases*. In the hierarchical subdivisions (54 in all), the three main categories of the mineral kingdom, *i.e.*

Petrae, *Minerae* and *Fossilia* provide a niche for none of them. One entry will suffice to show how little relation the hierarchy had to the composition of the substances they embraced:

II. MINERAE: 2. Sulphura:

(a) *Ambra;*   (b) *Succinum;*   (c) *Bitumen;*

(d) *Pyrites;*   (e) *Arsenicum.*

Before we turn to the possible influence of Linnaeus on scientific nomenclature in domains other than the biological sciences, we may pause to touch on the lighter side of the grand design. He starts his description of the *Regnum Animale* in vol. 1 of the *Systema*, with a classification of all legimate topics of human enquiry other than *Morborum* aforesaid. Besides NATURALIA (divisible into ANIMALIA and VEGETABILIA) and LAPIDES, the IMPERIUM NATURAE includes God, the celestial bodies, the four elements of the ancients (including *air*), Mankind, Wisdom and the Scientific Method (i.e. taxonomy). The inclusion of the *Imperator* himself in the foregoing list reminds us that Linnaeus was the devout son of a Lutheran pastor. For the benefit of the reader who has a smattering of Latin at his or her disposal, here are a few samples of the piety which pervades his *magnum opus*.

On the fly-leaf of vol. 1 of the 10th edition (1757) is the invocation: *O JEHOVA, quam ampla sunt Tua Opera, quam sapienter Ea fecisti, quam plena est Terra possessione Tua*. There follows on the back of the title-page: *Magnus est DEUS noster et magna est potentia Ejus et potentiae Ejus non est numerus*. The next two pages, listing previous editions and authorities, conclude with the prayer: *Docuisti me DEUS a juventute mea et usque nunc pronunciabo Mirabilia Tua*. The second paragraph of the *Introitus* (foreword) ends thus: *Finis* Creationis *telluris est gloria* DEI, *ex Opere* Naturae *per* Hominem *solum*. It concludes with: *O JEHOVA! Quam magnifica sunt Tua Opera! Vir incipiens non cognoscit ea et stultus non animadvertit ea*. The general discussion of the *Imperium Naturae* (pages 5–8) which precedes the outline of the *Regnum Animale* ends with a second citation from the Psalms: *Narrabo mirabilia Tua DOMINE et virtutem Terribilium Tuorum dicant generationes*.

The outline of the *Regnum Animale* (pages 9–13) concludes with the reflection *Terribilia sunt opera Tua DOMINE, in multitudine virtutis Tuae, Te metientur inimici Tui*. The last page (821) ends with a text from the Apocrypha: *Pauca haec vidimus operum DEI, Multa abscondita sunt majora his*. The back of the fly-leaf of each of the two succeeding volumes (*Regnum Vegetabile* and *Regnum Lapideum*) exhibits an invocation to Jehova, and the introductory glossary of vol. 2 (10th edition 1759) ends with an invocation from the Psalms. Its final page carries a text from the Apocrypha. The tenth edition (1768) of vol. 3 is more reticent till we reach (p. 32) the mid-page inscription *Semitae DEI in Abysso* (the footpaths of God in primordial chaos). It ends with a pious reflection with no explicit mention of the deity.

3 Background to the French Reform

Even more than the date of publication of the first edition of the *Systema Naturae*, the year 1787 signalizes a momentous contribution to the world-wide vocabulary of Western science. It was then that four French academicians published the *Méthode de Nomenclature Chimique*. The initiative had come from Guyton de Morveau. His associates were Lavoisier, Berthollet and Fourcroy. Though the primary concern of the four had a more limited scope, a memoir presented to the Academy by Lavoisier in justification of the project expounded the principles invoked as part of a more comprehensive programme of linguistic reform. Languages, he declared, 'are not merely passive signs to express thought, they are also analytical systems by means of which we advance from the known to the unknown and to a certain extent in the manner of mathematics. . . . A moment's reflection readily shows us that algebra is a real language; like all languages it has its representative signs, its method, its grammar, if I may use this expression. Thus an analytical method is a language and a language is an analytical method and these two expressions are, in some sense, synonymous.'

Continuing in the same vein, Lavoisier[1] insists:

if languages are really instruments fashioned by men to make thinking easier, they should be of the best kind; and to strive to perfect them is indeed to work for the advancement of science . . . this method which must be introduced into the teaching of chemistry is closely connected with the reform of its nomenclature. A well-composed language adapted to the natural and successive order of ideas will bring in its train a necessary and immediate revolution in the method of teaching . . . we shall have three things to distinguish in every physical science: the series of facts that constitute the science, the ideas that call the facts to mind and the words that express them. The word should give birth to the idea; the idea should depict the fact.

In the same memoir, Lavoisier mentions examples of contemporary chemical terms which demand 'much practice and a great memory . . . to remember.' Such are *powder of algaroth, salt of alembroth, pompholix, phagedenic water, turbrith mineral, aethiops, colchocar, oil of tartar per*

[1] Cited from Douglas McKie, *Antoine Lavoisier* (1952).

deliquium, oil of vitriol, butter of arsenic, butter of antimony, flowers of zinc.
The translator of the second edition of the *Méthode de Nomenclature Chimique* mentions others with their reformed designation; *blue vitriol* (copper sulphate), *Rochelle Salt* (potassium sodium tartarate), *salt of wormwood* (potassium tartarate), *tartar emetic* (potassium antimony tartarate).

In his youth, Lavoisier had undertaken a mineralogical survey; and it is likely that the *Regnum Minerale* of Linnaeus first turned his thoughts to the chaos of chemical nomenclature at that time. Be that as it may, one can scarcely doubt that the scope of the third volume of the *Systema Naturae* would have been different if the science of chemistry had been as flourishing, and the prospects of chemical industry as promising, when Linnaeus published the first edition, as they were when he died. His death occurred in 1778 towards the end of a decade which witnessed prodigious advances in the study of the gaseous state.

Apart from the quantitative work of Black (1756) on the composition of carbonates, one can record little progress in exploration of the gaseous state between 1700 and 1770. A period of stagnation succeeded what had been a promising start during the second half of the seventeenth century. Before 1650, Van Helmont had shown how to prepare *gas sylvestre* (CO_2). In the *Sceptical Chymist* (1661) Boyle had shown himself to be aware of the existence of a type of inflammable air (i.e. hydrogen) different from the coal gas of the mines. Before 1670, the work of Hooke and Mayow had laid the foundations for a scientific treatment of combustion and respiration by showing that: (*a*) air consists of two sorts of *particles*; (*b*) one sort (which we now call oxygen) is essential to respiration and combustion; (*c*) the other, and inert, sort (now called nitrogen) cannot support combustion or respiration; (*d*) the gas (i.e. O_2) obtained by heating nitre (KNO_3) restores to air exhausted by combustion its power to sustain it.

A century later, in the decade during which Linnaeus died, discovery of other gases followed in quick succession. In England, Priestley (1772–4) discovered *inter alia* nitrous oxide, ammonia, sulphur dioxide and hydrochloric gas. In Sweden, Scheele (1774) isolated chlorine. Two years later, Cavendish established that hydrogen is both an element and a constituent of water. Meanwhile, in France, De Lassone discovered carbon monoxide, and before the end of the decade Priestley, Scheele and Lavoisier resumed, where Hooke and Mayow had left it, the study of the role of oxygen in combustion. By 1780 a drastic overhaul of current concepts and of nomenclature was thus overdue.

The efflorescence of chemical discovery in Britain, France and Sweden during the second half of the eighteenth century coincided in Britain itself with an industrial revolution, one facet of which has been grossly neglected by economic historians other than Nef. The exhaustion of wood as a source of industrial fuel and its replacement by coal in the latter part of the eighteenth century encouraged the search for substitutes of the many by-products of wood: charcoal for making gunpowder and for reducing metallic ores, alkali (*potashes*) for glass and soap manufacture, tar for ships' bottoms. Partly for this reason and partly because new machinery, which greatly accelerated production in general, created new markets for substances used by manufacturers, chemical industry of a new type and on a new scale was also emerging.

What chemical industry existed at the time when the *Systema Naturae* first appeared was not impressive. To do justice to its author's neglect of what we now call chemistry, one should bear in mind that few of the *pure* substances which are its main concern occur abundantly in nature, that freshly fallen rain-water is indeed the only one of them found everywhere and that industry in the first half of the eighteenth century still relied heavily for raw materials on such crude ingredients as sand, clay, potashes (then called alkali), chalk, sea salt, lard or whale oil and bone ash. Chalk used in the kilns of the brick fields to make quicklime for cement might have a modest claim to rank as a chemical entity; but the product evaporated in the salt-pans from sea-water was by no means pure $NaCl$. It owed its usefulness as a desiccating agent to the fact that it contained over 10 per cent magnesium chloride. The alkali (potassium or sodium carbonate), required more especially for soap making, was obtained by incinerating wood charcoal or seaweed.

Two British industries had requirements for purified products before 1700. To make gunpowder one needed, besides a supply of charcoal from the charcoal burners and relatively pure imported sulphur, nitre recrystallized from the crude saltpetre of manure heaps. British textile manufacturers imported alum (usually double aluminium potassium sulphate) for mordanting before use of dyes. Before 1650, they had also used oil of vitriol (sulphuric acid) for bleaching, preparing it by dry distillation of green vitriol (ferrous sulphate), of which Britain had natural deposits on the Isle of Sheppey.

As a practising chemist, in contradistinction to his role as a herbalist, the eighteenth-century apothecary depended largely on a market of hypochondriacs for disposal of the products of distillation and crystallization. He could expect brisk business among the constipated by prescribing calomel (mercurous chloride), Glauber's salt (sodium sulphate) and Epsom salt (magnesium sulphate). His non-medical clientele included painters for pigments, e.g. vermilion (mercuric chloride), and jewellers who used sulphuric acid with nitre for separating silver from copper and a mixture (*aqua fortis*) of nitric with hydrochloric acid to remove base metals from gold.

Of all his commercial wares, sulphuric acid turned out to be the most rewarding. In 1736, Ward, a London apothecary, took out a patent for its manufacture by a process already used on the Continent, and started production in partnership at Twickenham. The time was propitious. Contemporary expansion of the Scottish linen industry offered good prospects for use of the acid as a bleaching agent. Ten years later Roebuck, a Scottish medical graduate, went into business in Birmingham to produce it by a similar process, i.e. with sulphur and nitre as raw materials; but Roebuck's procedure had a novel feature which made production on a scale much larger than hitherto feasible. He substituted lead chambers for glass containers. Owing to legal difficulties about patent rights, he transferred his factory to Scotland in 1749.

Such was the beginning of large-scale chemical industry as we now think of it. It dictated its subsequent course. Without a considerable purchasable supply of sulphuric acid to generate the gas by reaction with zinc, the first ascent of a manned hydrogen balloon could not have taken place at a date so early as 1783. By 1797 there were at least six factories for making the acid in Glasgow alone. Sulphuric acid had long been used by apothecaries as an essential ingredient for converting sea salt into the *sal mirabile* of Glauber, and Glauber's salt roasted with coke and lime was an essential ingredient of the Leblanc patent for synthetic alkali (1791), i.e. sodium carbonate. Before Leblanc, James Keir, a Scots physician, and his partners started (1780) a factory at Tipton near Birmingham for the production of sodium hydroxide, using slaked lime and Glauber's salt as raw materials. Their firm also produced white and red lead for the Potteries. A by-product of making Glauber's salt was *marine* (hydrochloric) acid. When treated with manganese dioxide this yields chlorine, used briefly (1780 onwards) in Scotland for bleaching before Tennant of Glasgow embarked on commercial production of bleaching powder (1799) through its action on lime. Before 1792 Lord Dundonald had started in Scotland large-scale production of sal ammoniac (ammonium chloride), much in demand by dyers, braziers and tin-platers. The demand for dyeing derives from the earlier use of stale urine, which is rich in ammonia.

The foregoing sketch is by no means an exhaustive list of chemical industries which came into being between 1750 and 1800. For an authoritative and comprehensive account the reader may turn to *The Chemical Revolution* by A. and N. Clow. That their book deals so largely with the Scottish scene puts the spotlight on its relevance to our main concern. Britain was leading the world in emergent large-scale chemical industry during a period when Britain shared with France pre-eminence in chemical discovery and French savants were in the vanguard of a theoretical reappraisal which demanded a drastic over-haul of its vocabulary.

In short, the Chemical Revolution was well under way when Guyton de Morveau published (1781–2) a memoir on chemical termi-nology. According to Pearson in an introduction to the second (1799) British edition of the *Méthode de Nomenclature Chimique*, Guyton de Morveau explicitly formulated in the following terms principles which should guide a programme of reform:

1. Every substance should have a name in contradistinction to a phrase.
2. Names should be given 'according to the nature of the things intended to be signified by them.'
3. 'When the character of the substance is not sufficiently known to determine the denomination, a name which has no meaning' is better than one which conveys 'an erroneous idea.'
4. For 'new denominations, those which have their roots in the most generally known dead languages' are preferable, so 'that the word may be suggested by the sense and the sense by the word.'
5. The denominations should be consonant with the structure of the natural language which accommodates them.

After publication of the memoir of 1781, Guyton de Morveau continued, says Pearson, 'his labours to improve the language of Chemistry, and sensible of the extreme difficulty of rendering it perfect, wisely in 1787 availed himself of the assistance of the Members of the French Academy.' The enlistment of Lavoisier as a collaborator was especially fortunate. Other than the author of *Reflexions* on phlogi-ston (1783), no one then living could have contributed so much to the undertaking. Though the intention of de Morveau clearly embraced the possibility that a single name (compounded of two or more roots from a *dead* language) should have some relevance to the composition of a pure substance, there was still a formidable obstacle to its realization at the time when his memoir appeared. It has left its impress on the baptismal name of chlorine, i.e. *dephlogisticated muriatic* (brine) *acid.*

No advance towards a satisfactory interpretation of chemical composition was realizable till chemists themselves fully understood the role of oxygen in combustion and the formation of metallic oxides. Priestley and Scheele, often spoken of as discoverers of oxygen, in fact added little to what Hooke and Mayow found out. To be sure, they did *isolate* oxygen; but what they did, in this sense, discover they interpreted erroneously in terms of prevalent superstition. That Lavoisier was able to refute it by gravimetric methods set the stage for a rational interpretation of chemical reactions in general, whence also the possibility of embodying information about composition in the names given to compounds.

If the remarkable tempo of chemical discovery between 1770 and 1780 is less easy to explain, there is little doubt about what was a major circumstance contributory to the stagnation of chemical science between 1700 and 1770. In the closing years of the seventeenth, and throughout the first half of the succeeding century, Teutonic mysticism revived both Aristotelian teaching that fire is an element and Aristotle's discredited theory of gravitation by endowing a suppositious entity named *phlogiston* (from the Greek word $\phi\lambda o\xi$, $\phi\lambda o\gamma$- – for *flame*) with the peculiar properties of: (*a*) inflammability which escapes when things burn; (*b*) levity (negative weight) which makes hot air rise.

Between 1772 and 1777, Lavoisier and his compatriot Pierre Bayen carried out extensive experiments involving precise gravimetric measurements of the components in reactions involving oxidation, i.e. combustion and calcination of metals. One such experiment sufficed to discredit the current view. If true, gain of weight by conversion of a metal to its *calx* (oxide) is due to expulsion of phlogiston in contradistinction to removal of an atmospheric constituent. Lavoisier showed that calcination of tin by heating with air in a closed vessel results in a gain by the metal o ¦solid weight equivalent to increase of total weight by readmission of air, thus proving that the metal gained weight only by removing part of the air itself.

When Guyton de Morveau published his own memoir, Lavoisier had not completed the researches which undermined the last stronghold of Aristotelian superstition; but when the partnership between them and two other colleagues of the French Academy took shape, he had matured his views on chemical combination in the light both of his own researches and of predecessors who had, like Black of Glasgow, used gravimetric methods to put the study of chemical combination

33

on a quantitative as well as a qualitative basis. When the new nomenclature designated one oxide as *nitrogen trioxide* and another as *nitrogen pentoxide* the two labels thus convey:

(*a*) qualitative information that each is a compound composed of the two elements nitrogen and oxygen;

(*b*) quantitative information about the volumes in which they combine to make one or the other, whence also (from tables of their densities at a given temperature) their combining weights.

Approaching the joint enterprise as a linguistic innovation from a viewpoint other than that which Lavoisier adopts in the memoir cited at the beginning of this chapter, let us now focus attention on two of the five principles which Guyton de Morveau himself laid down, as summarized by Pearson. One is that 'those which have their roots in the most generally known *dead* languages are preferable for new denominations.' The other is that the denominations should be consonant with the *structure of the natural language which accommodates them.*

The first of these two requirements is *en rapport* with increasing reliance on Greek by biologists who followed in the footsteps of Linnaeus; and it bears the imprint of his influence on the vocabulary of science. The second, which refers to the structure of the natural language, leaves the way clear for an innovation which biological taxonomy did not anticipate. Besides using numerals such as *penta-* (Greek) or *sesqui-* (Latin) to convey quantitative information 'by a name in contradistinction to a phrase,' the taxonomy of the 1787 joint memoir implemented the recommendation that names should convey information about composition by giving a similar role to current suffixes of the *French* language. Because of our hybrid heritage, and because English is almost free of flexions such as those which encumber German, every relevant French terminal (*see* table on p. 13) is adaptable with only a trifling change of spelling to use throughout the English speech community. In expounding the role of the suffixes, Pearson himself seems scarcely aware of transition from French to English when he writes[1] in the introduction to his translations of the *Méthode de Nomenclature Chimique*, twelve years after its first publication, as follows:

A single example, taken from the New System of Chemistry, may explain clearly how much brevity and simplicity in terms, provided the terms have a

[1] Typographical conventions (caps and italics), as in the original.

proper import, facilitate the acquisition, retention and communication of chemical knowledge. SULPHUR may be a component of a great number ... of substances. ... SULPHUR may unite with OXYGEN by which combination it is rendered into the ACID STATE; but this ACID is of three different species, according to three quantities of Oxygen which may combine with a given quantity of Sulphur; and these three species are named the *sulphu*REOUS, the *Sulphu*RIC and the OXYGENATED *Sulphu*RIC acids. ... Each of these acids may unite with at least twenty-six different kinds of substances, which are metallic Oxides, Earths and Alkalis, and consequently produce seventy-eight different compound bodies. ... Accordingly, the word *Sul*PHITE denotes compounds consisting of the *Sulphu*REOUS Acid and each of the above twenty-six different kinds of substances; *Sul*PHATE implies compounds consisting of *Sulphu*RIC Acid and each of the above twenty-six ... and ... *Oxysul*PHATE signifies compounds consisting of the OXYGENATED SULPHURIC Acid and the above twenty-six. ... The particular species of compound substances belonging to each of these genera, named *Sul*PHITE, *Sul*PHATE and *Oxysul*PHATE, are signified by subjoining the name of the basis as an adjective to these generic names. Accordingly, the meaning of the names *Sul*PHITE of Soda, *Sul*PHATE of Soda, *Oxysul*PHATE of Soda will, without difficulty, suggest the composition of these substances. ... SULPHUR may unite with metals, Earths, Alkalis, Hydrogen Gas and other bases which are not acidified or are not acidifiable. The compound bodies produced by these combinations are denominated SULPHURETS[1]. ... By the names SULPHURET OF POTASH ... OF SODA ... OF IRON ... OF LIME ... OF HYDROGEN GAS, etc., a just notion may be acquired of the composition of the compound just mentioned to consist of SULPHUR AND CERTAIN BASES NOT ACIDIFIED or NOT ACIDIFIABLE (pp. 4–7). ... It appears, therefore, that by four different terminations of the word Sulphur and two different abbreviations of it, and by adding the word which is the name of the species of substance combined with a compound of Sulphur, Oxygen and a basis, above 300 different kinds of substances (which consist of Sulphur united to other bodies) may be denominated, so as to import the most essential properties of the things which these terms are intended to signify (p. 8).

If the English translator did not feel the need to comment on any minor adjustments of this aspect of the new taxonomy to the structure of his own vernacular, it did not escape his attention that a proposal made with little prevision of its international acceptability could not in Germany earn a reception as cordial as in Britain. The system of suffixes English had inherited from Norman French or, like French from classical Latin, was both alien to German and difficult to adapt to a

[1] Now *sulphides*, in agreement with *oxide*, which is the new nomenclature adopted from the start. The affix -*ide* is a classical assimilation from Greek εἶδος for *species* in contradistinction to *genus*.

language whose adjectives and nouns have different case forms singular and plural. Pearson remarks somewhat naïvely (p. 17 – *italics inserted*):

> in Germany, Gulanner appears to have been the first who introduced the new chemical nomenclature of the French; but in rendering it into the German language, he has made *several changes in the words of the names suitably to the nature of the language*. . . . As instances may be mentioned, Salpetergesaure Pottasche, Salpetergesaurte Pottasche, ueber saure Saltpetergesaurte Pottasche, for nitrite, nitrate and oxynitrate of Potash.

While Lavoisier was able to take advantage of the fifth of the principles which Guyton de Morveau had laid down, it thus seems clear that the latter was thinking of reform in national terms with no anticipation that 'the structure of the national (i.e. French) language' might create a formidable obstacle to acceptability for international use. The hybrid character of the English vocabulary thus played a decisive role in promoting the acceptance of the French reforms in countries other than France. Prompt acceptance of the new nomenclature by Britain, at a time when Britain and France were in the vanguard of chemical discovery and chemical industry, confronted the international scene with a *fait accompli*. It did so by the unforeseen accident that two national languages had the same battery of suffixes.

Chemical nomenclature did not circumscribe the contribution of Lavoisier and his colleagues to linguistic reform. They also turned their attention to the designations of units of measurement. At the outbreak of the Revolution there was no single system of weights and measures for France as a whole. When the French National Assembly (1790–1) appointed a commission of savants to make recommendations with reference to a uniform system of weights and measures throughout the country, Lavoisier became its first secretary and treasurer. Like the reformers of chemical nomenclature, the commissioners aimed at defining units in meaningful and memorizable terms. To be sure, they were powerless to change the unit of time then adopted by astronomers, cartographers and sea captains of the Western world, i.e. the second, $(86,400)^{-1}$ of a mean solar day. Otherwise, they were free to legislate for the future.

The Swedish astronomer Celsius had already (1742) defined a decimal unit (*degree*) of temperature with a simple relation to the thermal properties of water, 0° being the freezing and 100° the boiling-point of water at sea-level. The commission decided to define a unit of length (*metre*) in decimal terms equally memorizable. They defined the metre as one ten-millionth of a half meridian of the earth's surface. To

36

determine this with the utmost accuracy, they set on foot a geodesical investigation which continued for several years before completion. The National Assembly adopted it in 1799. Five years before that, Lavoisier had become one of the casualties of the Revolution; but the commission remained steadfast to Lavoisier's plea that 'the word should give birth to the idea.' A unit of volume (*litre*) being 1,000 cubic centimetres, they defined 1,000 grams (a *kilogram*) as the mass of a litre of water at its maximum density (i.e. at 4°C).

The way was now clear to define (at a much later date) the *calorie* in equally meaningful and memorizable terms as the amount of heat required to raise 1 gram of water through 1° centigrade, the *dyne* as the force required to impart to a mass of 1 gram an acceleration of 1 centimetre per sec² and the *erg* as the amount of work done by 1 gram moving through 1 cm with uniform acceleration 1 cm². The last three examples, like all the recommendations of the French Commission (1791–9), adhere to the rule of choosing roots from the 'most generally known dead languages'; but despite the lead given by the French architects of the C.G.S. system, physicists and engineers, long accustomed to such vernacular terms as *force*, *speed* and *work*, were slow to change their ways.

In the perennial warfare between salesmanship and scholarship, commercial horticulture had staked a claim for use of *eponymous* terms before international conferences in the latter half of the nineteenth century reached agreement on the definition of electrical units. Eponymous terms are terms based on the name of a discoverer, innovator or patron, and this *genre* of word-building had a peculiar fascination for nineteenth-century physicists. Botanical examples are *Thunbergia*, a plant genus, and *Berberis thunbergii*, a plant species, both named as such after the botanist Thunberg. Sometimes the outcome is slightly comic, e.g. *Kniphofia*, the generic name of a plant whose inflorescence has a ruddy phallic aspect, whence called *red-hot poker* in the vernacular. Early in our own century, an American palaeontologist proposed the creation of a species *Hesperopithecus harold j cookii* to accommodate the discovery of a molar tooth by one Harold J. Cook.

This backward step in biological nomenclature set the fashion for combing obituary notices of departed worthies to upholster names of physical units such as *ampere*, *angstrom*, *coulomb*, *curie*, *farad*, *fermi*, *gauss*, *henry*, *joule*, *newton*, *ohm*, *roentgen*, *volt* and *watt*. The only excuse for this outbreak of ancestor worship is the extreme difficulty of finding suitably suggestive Latin or Greek roots for types of measurements with no parallel in antiquity.

4 Spelling Conventions as Tools of Diagnosis

There will be times when a student of natural science or a professional scientific worker will wish to trace the origin of the roots which make up a technical term or, it may be, to invent a new one. It might seem that the only requirements are a couple of two-way dictionaries, one Latin, the other Greek, and a knowledge of the Roman alphabetic signs equivalent to the Greek letters. For several reasons, this view is unduly optimistic.

Owing to the fact that contemporary school instruction in English-speaking countries contains no niche for the systematic study of etymology, few readers of this book will realize one great merit of our much abused Anglo-American orthography. Far more often than not, it is possible to distinguish with the help of a few rules whether a word is of native (i.e. Anglo-Saxon) origin, of Greek origin or of Latin origin, and, if Latin, whether from classical sources or modified through the medium of Norman and Angevin usage. Only rules for distinguishing Greek from Latin words by their spelling concern us in this context, and some readers would be content with a brief list. They will find it on pp. 44–46. Others will be curious about their rationale and, for such, a brief sketch of the history of the alphabet is necessary to fill in a lacuna of early education.

All alphabets have a common origin.[1] Writing by recourse to their symbols began nearly four thousand years ago among Semitic people with access to Egyptian quasi-pictorial writing. From Egyptian hieroglyphs which stood for whole words, they borrowed signs for sound values, regardless of their original meaning. The maritime Semitic trading rivals of the Greeks spread alphabetic writing throughout the Mediterranean. There Greek-speaking island colonies adapted it (about 600 B.C.) to their own linguistic requirements, transmitting it to the Etruscans and to other peoples of Italy. Though the common ancestry of the classical Greek and Roman alphabets is transparent, divergences of sound values and the signs of other alphabets makes it difficult to detect their origin from one source.

[1] Except indirectly, the cryptographic Ogam script of Celtic inscriptions.

The several reasons for this explain one or other of the differences between the Greek and the Roman. These are:

(i) The earliest use of alphabetic sound-writing was for short inscriptions written with no fixed order of the signs horizontally or vertically. When need for order asserted its claims, writing from left to right or vice versa and top to bottom or vice versa led to giving the same primitive sign a different vertical or horizontal orientation.

(ii) The writing surface (stone, wax, papyrus, parchment) and the appropriate writing tool (chisel, stylus, brush, quill) influenced whether the final shape should be angular or curved.

(iii) Languages of communities which cannot freely communicate have evolved in historic times from intercommunicable dialects. In this process, some sounds have locally displaced others. So the sign referable to a particular sound at one stage in the life-history of a speech community, comes to stand for another at a later stage.

(iv) When one speech community transmits the art of alphabetic writing to another, some signs may therefore be redundant and for some sounds the parent alphabet may offer no signs. One way of making the best of a bad job is then to give redundant signs a new sound value.

In short, identical signs or signs with the same parentage may stand for different sounds in different speech communities or even (as in English) in the written word of the same speech community. For this reason, and because many readers will not be familiar with the International Phonetic Alphabet now used by philologists, it will here be convenient to use capitals for signs, e.g. C in *coat*, and small letters for sounds, e.g. *k* as in *kit* for C as in *coat*.

As regards sound shifts mentioned in (iii) above, the reader with a smattering of German will be aware of two which distinguish High German from its sister Teutonic languages, including Old English. One is the substitution of *ts* (= Z) for *t* initially (e.g. *Zunge* = *tongue*) or *s* (= SS) for *t* elsewhere (e.g. *Wasser* = *water*). The other is the substitution of *pf* for *p* initially and otherwise *f* for *p*, as in *Pfeffer* = *pepper*. A very common class of word shifts even among intercommunicable dialects is from voiced to unvoiced partners (or vice versa) of such pairs as the following: *b – p, d – t, g – k, v – f, z – s*. In this sequence, the first member is the voiced one, *g* and *s* having their characteristic values as in *god* and *sod*.

The range of sounds both of Caesar's Latin and of Aristotle's Greek was narrower than that of modern English. Neither embraced the sounds represented by:

(i) SH in *ship* and CH in *chef* (∫ in the International Phonetic Script);

(ii) S in *measure*, SI in *fusion*, J and G in French *juge* (ʒ in I.P.S.);

(iii) TH in *then* (*not* as in *thin*) ð in I.P.S.;

(iv) Y in *year* and J in German *Jahr* (j in I.P.S.);

(v) J in *June* (dʒ in I.P.S.);

(vi) TCH in *catch* (t∫ in I.P.S.).

Caesar's Latin did not accommodate the sound (*v*) represented by V in *vain* and F in *of*, by Z in *zebra* and S in *his*, by TH in *thin* and by NG (ɲ in I.P.S.) as in *sing*. Aristotle's Greek embraced neither the *v*-sound nor the *f*-sound represented by F in *fit* or FF in *off*. Nor did it accommodate the sound represented by W in *woe* or *wit*. Classical Greek did, however, embrace two sounds alien to Anglo-American, albeit one occurs in Scots dialect, and both in Old English. The sign χ stands for the *aspirated k-sound* in *loch* as a Scotsman pronounces the word. The Greek sound represented in Roman script by RH existed in Old English, written by Anglo-Saxon scribes as HR.

Inscriptions disclose many variants of the earliest alphabetic signs adapted from Semitic sources to their own requirements by traders speaking different Greek dialects in colonies extending from the coast of Asia Minor to what is now the south of Spain. We can thus reconstruct how the Greek alphabet, as we now know it, took shape in

Greek			Roman	Greek			Roman
A	α	*alpha*	A	*N*	ν	*nu*	N
B	β	*beta*	B	*Ξ*	ξ	*xi*	X
Γ	γ	*gamma*	G	*O*	ο	*omicron*	O (short)
Δ	δ	*delta*	D	*Π*	π	*pi*	P
E	ε	*epsilon*	E (short)	....			Q
....			F	*P*	ρ	*rho*	R
Z	ζ	*zeta*	....	¹*Σ*	σ/ς	*sigma*	S
H	η	*eta*	E (long)	*T*	τ	*tau*	T
Θ	θ	*theta*	(TH)	*Y*	υ	*upsilon*	U *or* V
I	ι	*iota*	I	*Φ*	φ	*phi*	(PH)
K	κ	*kappa*	C	*X*	χ	*chi*	(CH=kh)
Λ	λ	*lambda*	L	*Ψ*	ψ	*psi*	(PS)
M	μ	*mu*	M	*Ω*	ω	*omega*	O (long)

¹ ς at the end of a word – otherwise σ.

times when the sound values of the same signs were not universally identical. We can also trace the ancestry of the Roman alphabet from humble beginnings when there was as yet no firmly fixed orientation of the individual signs. How far Greeks who then occupied the south of Italy directly contributed to it and how far indirectly through their Etruscan pupils is uncertain. When comparing the Roman with the Greek alphabet, one should bear in mind that the Greek capital letters constitute an earlier stratum than their small (so-called *minuscule*) partners, being in fact closer to forms used for inscriptions before there was a literature.

Before the conquest of Greece and its island colonies, Romans did not use all the symbols included in what we customarily call the Roman alphabet. The letters K, Y and Z were borrowed by later Roman authors from the Greek. Two others, J and W, are mediaeval artefacts. Before the break-up of the Roman Empire, C stood only for the *k*-sound in *coat*, G for the *g*-sound in *goat*, Q (always followed, as still, by U) for the *kw*-sound in *quote* – never for the *k*-sound in *oblique*. The main differences between the Greek alphabet of Aristotle and the Roman alphabet of Julius Caesar three hundred years later are as follows:

(i) Possibly because early Latin had no *voiced* equivalent (*g*-sound) to the Greek (unvoiced) *k*-sound, Γ (*gamma*), locally written sideways as <, came to represent the latter. Where rounded off to a shape more convenient for inscription on wax tablets, it became *C*. If this surmise is correct there was no need to take over the Greek *k*. In the third century B.C. the Romans, whose speech did at least by then employ the voiced guttural, added a tick to distinguish it (as G) from the symbol for the voiceless one.

(ii) At a very early date, the Greek-speaking communities dropped the sign F (*digamma*) originally used for the *w*- and locally, it is possible, by a very common phonetic shift, for the *v*-sound. Neither of these sounds exists in the spoken language of historic times. The Etruscans retained *digamma* as F for the *f*-sound (or for its voiced partner as in *of*). From them, the Latins may have taken F for its unvoiced value in *off*.

(iii) Early Greek communities used a Semitic sign φ (*koppa*) which they eventually dropped as they dropped *digamma*; but Latin retained it as Q.

(iv) Latin did not retain the Greek symbols θ, ϕ and ψ.

(v) Before standardization of its alphabet, some western communities of the Greek-speaking world seem to have employed X-χ as in Latin with the sound value *ks* of X. Classical Greek script, that of

Athens, retained it with the sound value *kh* and adopted from the usage of other dialects the sign *Ξ-ξ* for *ks*. The reason for the sound value of *χ* in the classical Greek alphabet is uncertain. It is however suggestive that the *h*-sound in many Greek words replaces the *s*-sound of Latin and of most other Indo-European languages in words of common ancestry, as exemplified by the meaning of equivalent roots in the following current technical terms:

hemichord	semiquaver
hexagon	sextant
heptagon	septuagenarian
heliograph	solar
halogen	saline
hypertonic	supersonic
hydroponic	sudorific

(vi) The Greek sign *Y*, written as *v* in some earlier inscriptions, stood for the French *u* in *lune*. In Caesar's Latin V and U were interchangeable ways of writing what was originally the same sign, the first more suited to chiselling on stone, the other to writing on wax tablets or papyrus. Either could stand for two sounds:

(*a*) before a consonant, a vowel sound represented by OO in English, OU in French;

(*b*) before a vowel the consonant sound represented centuries later in English by W (at first written UU, hence *double-U*) as in *woman*.

Before the end of the Empire, there seems to have been a tendency to restrict V for the consonantal *w*-sound and to use U only as a vowel symbol. By the time when the Western Empire began to fall apart provincial Latin dialects were breaking up into what we now call Italian, French, Spanish, etc., and the sound we now represent by *V* had already replaced the sound we now represent by *W*. Independently, this also happened in German and other members of the Teutonic family. In the modern descendants of Latin, as in English and in Scandinavian languages, V now stands for *v* in *vote* and U stands for a vowel sound; but English usage was not uniform till the eighteenth century.

(vii) The Roman symbols A, E, I, O, each had two values, long and short. Greek E and O were both short. The long *o* written as *Ω-ω* is probably an adaptation of O. The long *e* represented as *H-η* is a true descendant of the sign which represents the *h*-sound in the Roman alphabet. It may be that the *h*-sound had disappeared in some local Greek dialects, as in modern descendants of Latin. If so, it would be available for a different use. At a very much later date, there came into use a sign for so-called *rough breathing* (i.e. the *h*-sound). This is the mirror image of the apostrophe in *man's*. Thus the Greek equivalent of the word Romanized as *hemi* in *hemisphere* is ἥμι or (in caps) ῾HMI. Greek dictionaries do *not* list separately words beginning with an aspirated vowel. So one finds a word whose Latin equivalent begins with HA-, HE-, HI-, HO- and HOU-, under ἁ, ἑ or ἥ, ἱ, ὁ or ὡ and οὑ cheek by jowl with the unaspirated vowels written redundantly with apostrophe in reverse as ἀ, ἐ, ἠ, etc. In Romanized spelling of some Greek words we do not find the appropriate *H* as after αν-. This prefix equivalent to α- when the root which follows begins with a consonant (as in *agnathous*) is translatable as *without* (before a noun) or *not* (before an adjective). We meet αν- in *anorexia* (without appetite) and in *anaemia*. The second root of the latter is the first root of *haemoglobin* and *haematin*.

To use a Greek dictionary for finding the meaning of a root contained in a scientific term, one also needs to know the way in which Latin authors represented the corresponding word in their own symbols, i.e. those we ourselves use. The Romans rendered Greek K by C. The only inconsistency of Greek spelling which calls for care when using a lexicon is the use of γ before one of the gutturals γ, κ, χ. It then stands for the sound otherwise represented by NG (ŋ in I.P.S.) in *sing*. Thus we pronounce:

ἀγγελος (*messenger*) as in *angel*
ἀγκυλος (*curved*) as in *Ancylostoma*
ὀγχνη (*pear*) as in *Onchnesoma*

As regards vowel symbols, Roman E had to do service for both ε and η, Roman O for both ο and ω. For Greek noun terminals -ος, -η or -α, and -ον, the Romans substituted -US, -A, -UM. The Roman conventions for double Greek vowel symbols were as follows:

Greek	Latin	Example
αι	AE	αἱμα (*blood*) as in *haematin*
ει	I	εἰδωλον (*image*) as in *idolatory*
οι	OE	οἰσοφαγος (*oesophagus*)
ου	U	οὑρα (*tail*) as in *urostyle*
ευ	EU	ευ as in *eulogy*

With some exceptions British writers conform to Roman usage as in *oecumenical* from οἰκουμενικος; but *ecology* (of animal and plant house-mates) and *economics* (house management = stewardship) from οἰκος (*house*) are exceptions. American writers commonly use E where Roman usage dictates. AE, e.g. in *pediatrics* from παις, παιδος (child). Like many proposals for spelling reform, this introduces scope for confusion, since *pedi-* (Latin), as in *pedicure* (foot-care), is the internationally current root for *foot*.

We are now ready to sort out trademarks of Greek origin in roots which contribute to internationally current terms.

(i) *CH* (χ) when pronounced in Anglo-American like *k* as in *chorus*, *chaos* and *chasm*, in contradistinction to CH pronounced as in *church* or in *chef*.

EXAMPLES: chromium, chlorine, synchronism, Branchiostoma, Chilopoda, Chenopodiaceae, Chrysochloris.

(ii) *K* (κ); in some words we employ the Greek K, incorporated in the latter-day Latin alphabet, instead of the C of Caesar's Latin.

EXAMPLES: karyokinesis, kinetic, kymograph, kaliapsis.

N.B. The French savants who drafted the C.G.S. system of physical units adopted K in preference to KH or CH for χ in χιλιοι (= 1,000), whence *kilometer*, *kilogram*. Most zoological terms retain the CH convention.

(iii) *PH* (φ) standing for the *f*-sound as in *phonograph*. In Caesar's time φ stood for an aspirated *p*-sound as in *uphill*; but its pronunciation in spoken Greek had changed to *f* before the end of the Western Empire. Late Latin authors used F, as do Italians today (cf. Italian *filosofia* for French and German *philosophie*). Scandinavians follow Italian usage.

EXAMPLES: phthisis, phlebitis, phloem, phlegm, Phaeophyceae, Siphonoglyphe, Phlox.

(iv) *RH* (ῥ), pronounced in Anglo-American without the Greek *rough breathing* (aspiration), occurs only in roots of Greek origin.

EXAMPLES: diarrhoea, rheostat, rheumatism, Rhinoceros, Rhynchocephalia, Rhododendron.

(v) *TH* (θ), initially pronounced like TH in pothole but later as in *thin* (*not* as TH = ծ in *then*), is also a common ingredient of native English words; but none of these contributes to internationally current terms. In all such, it is a trademark of Greek origin.

EXAMPLES: thecodont, thyroid, thymus, asthenic, euthanasia, ophthalmic, Theromorpha, Thalassema.

(vi) *Y* (*Y*, *υ*), like *Z* (below) adopted after the Roman conquest of Greece, occurs initially and terminally in many native English words. None of these contributes to internationally current roots, in which the Anglo-American pronunciation is commonly as in *pyrex* and *lyre*.

EXAMPLES: glycol, dynamo, gyroscope, mycelium, hypha, zygote, pyrites, phylum, cyanide, Cyprus, Chlorophyceae.

N.B. In *glycerine*, *syndrome* and *Cypripedium*, the pronunciation of Y like I in *pit* is anomalous.

(vii) *Z* (*Z*, *ζ*) was pronounced in classical Greek as either *z* or *dz* but in Anglo-American usage always as *z* in *zephyr* and in *zebra* (a word of Central African origin). A few words beginning with Z are of Arabic origin, e.g. *zirconium*, *zero* and *zenith*. Otherwise it is a trademark of Greek.

EXAMPLES: zone, zygomatic, zodiac, zoophyte, enzyme, Zygaena, Zeuglodon, Zostera.

*　　*　　*

Besides the foregoing, certain compound consonantal sounds and corresponding symbols, though not indicative in another situation, are specifically Greek at the *beginning* of words contributory to international scientific terms.

(viii) *BD-* (*βδ-*) occurs commonly in only one internationally current root based on the word for a *leech* (*βδελλα*), as in many zoological names.

EXAMPLES: Bdellostoma, Acanthobdella, Bdellophis.

(ix) *CN-* (*κν-*) as an *initial* compound consonantal *sound* is common in Teutonic languages where now spelt KN and still pronounced as such in Chaucer's time. It occurs in several *eponymous* names of plant genera (*Knightia*, *Knoxia*, etc.), but Romanized as CN- only in roots of Greek origin.

EXAMPLES: cnidoblast, cnemial, Cnemidophorus, Cnethocampa.

(x) *CT-* (*κτ-*) as an initial combination is exclusively Greek, mainly based in *κτεις* (stem *κτεν-*) for comb.

EXAMPLES: ctenidium, ctenoid, Ctenophora.

(xi) *GN-* (*γν-*) as an initial combination is exclusively Greek, found in technical terms.

45

Examples: gnosis, gnomon, agnathous, Gnathostomata.
N.B. *Gnu* is a word of Hottentot origin.

(xii) *MN-* (μν-), like BN- (above) and PN- (below), occurs only in one widely current component of technical terms.

Examples: amnesia, mnemonic, mnemotechnic.
(xiii) *PN-* (πν-), pronounced as *n* if at the beginning of a word.

Examples: pneumatic, pneumonia, pneumatophore.
(xiv) *PS-* (ψ-), pronounced in Anglo-American as in *psalm* and *psychology*, is exclusively Greek as an *initial* combination.

Examples: psoriasis, psittacosis, pseudopodium, Psylla, Psilotum, Psammobatis.

(xv) *PT-* (πτ-), pronounced in Anglo-American as *t*, if at the beginning of a word, is also uniquely of Greek origin.

Examples: pterygia, Pteridophyta, Coleoptera, Pterodactyl, Archaeopteryx, Pteropus, Litopterna, Ptychodera.

(xvi) *X-* (ξ) originally stood for the sound *ks* and now pronounced in Anglo-American as *z*, e.g. in *xylophonist*. As such it is uniquely Greek.

Examples: xylem, xerophyte, xanthophyll, xiphisternum, Xenopus, Xiphias, Xenia.

*   *   *

There are very few diagnostic peculiarities of the Latin sound system or of classical Latin orthography. For reasons already stated, i.e. that classical Greek had neither the *f*-sound nor the *v*-sound, both present in *late* Latin. Thus F and V (with its value in mediaeval Latin) are trademarks of Latin ancestry. The only other noteworthy trademark of Latin origin is QU for the *kw*-sound, which did not exist in classical Greek. This sound, represented by Anglo-Saxon monks as CW, did occur in Old English, e.g. *cwen* for *queen* and *cwic* for *quick*. Norman scribes replaced CW by QU, but no native English words in which it occurs are relevant to our theme.

The small group of Semitic roots preserved from the golden age of Moslem science discloses few diagnostic indications of the sort here discussed. In many, the definite article *al* is the initial syllable: *alchemy*, *algebra*, *alkali*, *almanac*, *alizarine*, *alcohol*, *algorithm*, *Aldebaran*. However, AL- occurs as an initial syllable in Latin words, e.g. *albumen* and *albino* both based on the adjective *albus*, etc. (white).

5 A Few Hints about Latin and Greek Grammar

Modern terms for animal and plant species are *binomial*, as are not a few anatomical terms and some chemical terms which are survivors of the days before the French Reform of Chapter 3. Students sensitive to language will have noticed that the suffixes (terminals) of both components are often alike, as is true of *Equus caballus* (the domestic horse), *Aqua regia* and *Polygonum fagopyrum* (buckwheat). Sometimes they are different, as is true of *Corpus callosum, Taxus baccata* (the yew tree), *Polygonum Convolvulus*. To understand why this is so, and to name correctly a new species in accordance with the rules of the game, raises a grammatical issue. Readers of this book, all or almost all of whom will have no knowledge of Greek, and at best only a meagre smattering of Latin long since relegated to the limbo of lost scholastic luggage, can get to grips with this aspect of internationally current scientific nomenclature only if willing to refresh their memories, or to make first acquaintance with what grammarians call *gender concord*.

If he has no prior familiarity with the elements of Latin and Greek grammar, other difficulties, besides those mentioned in the last chapter, thus confront the Anglo-American student of science or the professional scientific worker who wishes to understand and to use the build-up of the world-wide vocabulary of science. These arise from the fact that both languages are more highly inflected than the contemporary ones more commonly taught in the schools of Britain, the U.S.A. and the British Commonwealth.

To say that a language is more or less highly flexional means that we are more or less dependent on a textbook of grammar before we can use a dictionary to full advantage. Since Anglo-American is by far the least inflected member of the Indo-European family, which includes *inter alia* Russian, Hindi, Urdu, Greek, Latin, German, French and Spanish, this statement may not be obvious to every reader of this book.

A few examples will suffice to illustrate what it means. If we look up *cats* or *loved* in the *Concise Oxford Dictionary* we shall find no corresponding entries. The editors assume the user to know that:

(*a*) *cats* is a flexional form of *cat*;
(*b*) *loved* (like *loves* and *loving*) is a flexional form of *love*.

So, too, we shall not find *men* or *gives* as a *first* entry, albeit we shall find:

(i) the plural form *men*, in brackets after *man*;
(ii) the two entries *gave* and *given*, in brackets after *give*.

Even so, we do not find listed the flexional forms *man's* and *men's* or *gives* and *giving*. The assumption implicit in the editorial decision to list a word as a first entry, to tuck it away as a second or to leave it out altogether is that there is no need to list plural forms of the immense majority of nouns to which we merely need to add -*S* to the singular (dictionary) form. Nor is there need to cite the forms *loves*, *loving*, *gives*, *giving* formed in the same way from the dictionary (so-called *infinitive* and main *present tense*) form of all verbs other than the auxiliaries (*be*, *have*, *may*, *can*, *must*, *shall*, *will*). In short, there are many words which a Chinese student would not be able to trace in an English dictionary without some premedication of the sort one calls grammar. An Anglo-American student needs such premedication before he or she can use a Latin–English or Greek–English dictionary to full advantage.

No difficulty arises when dealing with the *particles* (*prepositions*, *conjunctions*, *adverbs*), which are invariant, nor with the *pronouns*, only one of which (*ego* in Latin) = ἐγω (in Greek) is a formative root, as in *egocentric*. Obstacles confront us only when we need to deal with the classes of words we call *nouns*, *adjectives* and *verbs*. An English-speaking student who has a smattering of German will experience little difficulty in appreciating the vagaries of the first two; but the reader whose only second language is French or Spanish will not anticipate that nouns and adjectives have the different flexional forms distinguished by what grammarians call *case*.

All English adjectives are invariant and most English nouns have two forms (singular and plural) only. A few have a distinct so-called *genitive* case form distinguished from the plural on paper by an apostrophe (e.g. *day*, *day's*, *days*) but pronounced in the same way. A tiny class have distinctive genitive forms both in the singular and in the plural (e.g. *man–man's*, *men–men's*). No clear-cut thread of meaning connects the terminal -*S* of *man's* in *a man's dentures*, *a man's deeds*, *a man's debt* and *a man's death*. The best way of defining the *genitive singular* (*man's*) is to say that it is the form we replace by *his*. The *genitive plural* (*men's*) is the form we can replace by *their*. French and Spanish nouns and adjectives have no separate case forms. Like *house–houses*, the French and Spanish noun has only two forms;

but the adjective has four, owing to a type of flexion (*gender*) which had altogether disappeared in English by Chaucer's time.

Let us recall what gender means. Every French or Spanish noun is classifiable as: (*a*) *masculine*, if replaceable as verb subject by the equivalent of *he*; (*b*) *feminine*, if replaceable as verb subject by the equivalent of *she*. Otherwise, the distinction has little to do with sex. The adjective of French and Spanish has distinct masculine singular, masculine plural, feminine singular and feminine plural forms, of which the dictionary cites only the first if the other three conform to the dominant (so-called regular) pattern. The rule (*concord*) for using them is that the singular variant accompanies a singular noun, the plural variant goes with a plural noun, masculine with masculine, feminine with feminine.

This takes us a step nearer from the invariant adjective of Anglo-American to the vagaries of noun and adjective concord in Latin and Greek. Contemporary German takes us two steps further forward. German nouns are classifiable as masculine, feminine and neuter, the last being nouns replaceable as subject of the verb by *es*, the equivalent of *it*. All nouns have both singular and plural case forms, of which at most four of each number category are distinguishable. One case form is the genitive, corresponding to *man's* and *men's* above. The other three are those which we replace, if the noun is masculine and singular, in French by *il*, *le*, *lui* and in Spanish *el*, *lo*, *le*. One speaks of them respectively as *nominative*, *accusative* and *dative* case forms. The German adjective has separate case forms and separate gender forms both singular and plural. *Concord* (i.e. choice of the right form) implies that the adjective has to have the case and number form of the noun it qualifies, as well as the appropriate gender.

In general terms, the situation is almost exactly the same in German and in Greek. Like Latin also, Greek nouns are classifiable as masculine, feminine and neuter. In the singular, some Greek and some Latin nouns and adjectives have a fifth (*vocative*) case form for personal addresses, e.g. *Et tu Brute* for *Oh Brutus you too*, and there would be far fewer of them if the poets of antiquity had indulged less in addressing inanimate objects, as in *Land of Hope and Glory* or *Hail America*.

A considerable proportion of Latin nouns and adjectives have an additional *singular* case form, called the *ablative*, for use in a variety[1] of

---

[1] Even educationists most opposed to the use of the cane must concede the difficulty of conveying to schoolboys one such use – the so-called *Ablative Absolute* – without its aid.

situations. One is that it follows the particles *de* (= concerning) and *sine* (= without). For a sizeable minority of singular, as for all plural, nouns and adjectives one case form does the work of both the dative and the ablative when recognizably different. It is not therefore strange that the Romans, who borrowed their notions of grammar from the Greeks, did not recognize the ablative till Julius Caesar drew attention to it in a memoir written during his campaigns in Gaul. *En passant*, one may note that the usual textbook lay-out of *cases* suggests that every Latin noun and adjective has six case forms in the singular and six in the plural, i.e. twelve in all. Actually none has more than eight. Many have only six.

By the end of the Western Empire, case distinction had largely disappeared and the masculine gender class had engulfed the neuter. The singular case form which ousted all others in the daughter dialects which became Italian, Spanish and Portuguese was the ablative or the common dative–ablative. In the plural, the persistent case form was the accusative. This ends in *-as*, *-os* or *-es* in all Latin nouns except those of neuter gender; and the latter, as stated, came to conform with the masculine pattern, whence loss of neuter adjectival forms. A small minority of Old English nouns had the *-S* terminal in the plural. It became the dominant form in Anglo-American, owing to the impress of French after the Norman Conquest.

Where distinct from the dative, the glossaries of Part II cite the ablative singular case form for nouns and the masculine nominative singular for adjectives. Otherwise for nouns, the case form is the one which does the job of both dative and ablative. The intention is not primarily that this tallies with the practice of an Italian, Spanish or Portuguese dictionary. The main reason for doing so is that the singular *ablative* (or common ablative–dative) case form often tallies with the contribution which a Latin noun or adjective makes to a technical term, more closely than does the first entry in a Latin–English dictionary (i.e. *nominative singular*). A few examples will make this clear:

Nomin. Sing.	Ablat. Sing.	Meaning
FLOS	FLORE	flower (as in *floral*)
CORPUS	CORPORE	body (as in *corporeal*)
ORDO	ORDINE	row (as in *ordinal*)
SANGUIS	SANGUINE	blood (as in *consanguinity*)
LEO	LEONE	lion (as in *leonine*)
REX	REGE	king (as in *regal*)
CARDO	CARDINE	hinge (as in *cardinal*)

Sometimes, as when one speaks of a *corpus* of knowledge or the *axilla* (armpit), Anglo-American and/or international scientific nomenclature had absorbed the dictionary (*nomin. sing.*) form of Latin nouns and of the Romanized rendering of Greek ones. This is especially true of names for animals and plants, of which the *generic* designation (e.g. *canis* in *Canis familiaris* or *pyrus* in *Pyrus japonica*) is with few, if any, exceptions a nominative singular case form. Contrariwise, the *specific* component of the binomial nomenclature is commonly an adjective, as is *sapiens* in *Homo sapiens*. As such it must obey the law of *concord*, i.e. agree with its generic noun with respect to case, number and gender. The notable exception to this rule is one on which Linnaeus occasionally rings the changes, as when designating the type species of a genus by two names, the generic one being from *classical* (i.e. literary) Latin, specific one being the *vulgar* (i.e. colloquial) Latin equivalent, as in *Equus caballus* and *Felis cattus*. In such cases, the vulgar Latin is the one most likely to have survived in a descendant of Latin, e.g. for *horse* Spanish *caballo*, French *cheval*, and for *cat* Spanish *gato*, French *chat*. Though not the type species of a genus, *Lepus cuniculus* (cf. French *lapin*, Spanish *conejo*) is another example of such a combination.

Whether the specific name is adjectival or not, the same rules of concord apply. It is therefore fitting to give here a few hints about how far the terminals of the dictionary form of Latin and of Greek nouns indicate their gender. It is not possible to do so for any considerable majority of them; but a sizeable group is divisible as follows and a considerable class of adjectives each have corresponding gender, case and number terminals. They are as follows:

Class	Nomin. Sing.	Nomin. Plural	Ablat. Sing.
Masculine	-US	-I	-O
Feminine	-A	-AE	-A
Neuter	-UM	-A	-O

The corresponding classes of Greek nouns have for Latin nomin. -US and -I Greek -ος and -οι; for Latin -A and -AE, Greek -α or -η and; -αι, for Latin -UM and -A, Greek -ον and -α.

A small class of Latin neuter nouns (e.g. CORPUS, OPUS, CRUS, GENUS – nomin. plur. CORPORA, OPERA, CRURA, GENERA) do not fit in the -US pattern above, and names of trees with this nominative singular terminal are feminine. Thus it happens that the second (adjectival) component of the following names has the corresponding feminine terminal:

Pyrus aucuparia	Mountain Ash
Crataegus oxyacantha	Hawthorn
Fagus sylvatica	Beech
Mespilus germanica	Medlar
Quercus pedunculata and *Q. sessiliflora*	Oak

The gender of the generic does not dictate that of the specific name when the latter is a noun, e.g. *Dianthus Armeria*, as indicated by the initial capital letter. Generic names of some herbs are masculine with the -US terminal for noun and adjective alike, as in: *Carduus crispus, Carduus tuberosus, Carduus lanceolatus, Carduus Marianus*. The commonest generic terminal of plant names other than trees is -A and the adjectival specific usually takes the feminine form, i.e. -A if the adjective is one of the -US, -A, -UM class. Thus we have:

Fragaria vesca	Strawberry
Draba verna	Whitlow Grass

With generic plant names which end in -UM (or Greek -ON) the specific name usually takes a neuter terminal, e.g.

Epilobium hirsutum	Great Willow Herb
Conium maculatum	Hemlock (of Socrates)
Sedum reflexum	Reflex-leaf Sedum

When the adjectival component is not one of the -US, -A, -UM class, the neuter nominative singular terminal is often -E, as in

Sedum acre	*Yellow Sedum*

When names of animals involve both nouns of the -US, -A, -UM class and adjectives of the -US, -A, -UM class, usage is more consistent than with plant names. The following are typical:

Nautilus pompilus	Pearly Nautilus
Rana temporaria	British Common Frog
Distomum hepaticum	Liver Fluke

Whether Latinized or merely Romanized like *rhododendron*, which retains the Greek form of the terminal *-ov*, generic Greek names, like Latin ones, are usually nominative singular forms. And the nominative singular form may be the *final* root of a derivative compound name. Thus πους, Romanized as *pus* (*foot*), appears in *Bradypus* (sloth) and

52

Xenopus (clawed toad), but the stem of other case forms (i.e. ποδ-) in class names or other derivatives (e.g. *Hexapoda, Isopoda, Podostemon*). Similarly, the nominative singular θριξ, Romanized as *thrix* (*hair*), occurs in the generic name (*Aphilothrix*) of a gall wasp. In other compounds, the stem (τριχ-), Romanized as *trich-*, appears (e.g. *trichogyne*). Thus it is not the final component of the generic name *Trichomonas* (a flagellate).

Like -US, the nominative singular Latin noun terminal -ER is typically a trademark of masculine gender, and there exists a small class of adjectives whose nominative singular forms corresponding to -US, -A, -UM are -ER, -RA, -RUM. They include:

niger, nigra, nigrum	black
pulcher, pulchra, pulchrum	beautiful
tener, tenera, tenerum	tender

Examples:[1]

Lathyrus niger	black pea
Centaurea nigra	knapweed
Solanum nigrum	black nightshade

By comparison with *Lathyrus hirsutus*, *L. tuberosus*, *L. montanus* and *L. maritimus*, the initiated reader will deem *Lathyrus* to be masculine. Similarly with *Rumex* in

Rumex pulcher	fiddle dock
Rumex nemorosus	red-veined dock

There are, however, a few Latin adjectives whose singular nominative case forms follow the plan: -ER (m.), -IS (f.), -E (n.). Important are:

acer, acris, acre	sharp, acrid
celer, celeris, celere	swift

Examples:

Ranunculus acris	meadow crowfoot
Sedum acre	yellow sedum

Much larger than the *acer-celer* class of Latin adjectives is one of which the terminal -IS is common to both masculine and feminine nominative singular case forms, the corresponding neuter terminal

[1] All the binomial names of plants and their vernacular equivalents cited in this chapter follow the *British Flora* of Bentham and Hooker, 7th edition revised by Rendle.

being –E. This class includes several very common specific epithets of plant names, e.g.

agrestis	wild, rural	*perennis*	perennial
arvensis	field	*pratensis*	meadow
autumnalis	autumnal	*sylvestris*	woodland, forest
biennis	biennial	*vernalis*	spring
campestris	field, rural	*viridis*	green
muralis	wall	*vulgaris*	ordinary, common

officinalis   officially recognized by quacks as of medicinal use

Examples:

Ranunculus arvensis	corn buttercup
Trifolium arvense (n.)	hare's foot clover
Avena pratensis	wild oats
Geranium pratense (n.)	meadow crane's bill
Aquilegia vulgaris	columbine
Origanum vulgare (n.)	marjoram

The nominative singular form of so-called *present participles* (e.g. *repens*, creeping; *nutans*, nodding or swaying; *urens*, burning) ends in –ANS or –ENS for all three genders.

Examples:

Agropyrum repens (n.)	couch grass
Silene nutans (f.)	Nottingham catchfly
Lobelia urens	acrid lobelia

The terminal –US of a specific name attaches to a neuter nominative singular case form when the word itself is a Latin neuter comparative. The nominative masculine and feminine singular terminal of the comparative adjective is –IOR or –OR, as in *major* and *minor*. The corresponding neuter terminal is –IUS or –US.

Examples:

Scutellaria minor	lesser skull-cap
Vinca minor	lesser periwinkle
Thalictrum minus	lesser meadow rue
Polygonum minus	slender polygonum
Vinca major	greater periwinkle
Orobanche major	greater broomrape
Chelidonium majus	greater celandine
Antirrhinum majus	snapdragon

To the rule that -UM is the trademark of the neuter nominative singular of a Latin noun or adjective, only one *caveat* is mentionable. The genitive plural case form of all Latin nouns ends in -UM (usually preceded by AR-, OR-, I- or U-). This accounts for such rare anomalies as:

| *Sempervivum tectorum* | house leek (of the roofs) |
| *Stellaria nemorum* | wood stitchwort (i.e. stitchwort of the woods) |

One other troublesome exception to the rule that the specific component of a name in accordance with binomial nomenclature is usually a nominative singular adjectival case form whose gender agrees with that of its generic partner calls for further comment. Sometimes the specific name of a plant is a noun used for the generic designation of another species which it superficially resembles. This accounts for such exceptions as *Polygonum Convolvulus* and *Polygonum Persicaria*. The use of a capital letter to introduce the second component gives the reader a salutary warning. *Dianthus Armeria* (Deptford pink) is another example of this sort, *Armeria* being the generic name whose vernacular equivalent is *thrift*. A comparable class of anomalies arises when an author decides to transfer a species from one genus to another. Thus some botanists place the Lesser Celandine in a separate genus *Ficaria*, and others include it in *Ranunculus* as *R. Ficaria*, using the capital F to indicate that it falls out of step with the customary rule of concord, as in *R. hederaceus*, *R. bulbosus*, *R. flabellatus*, *R. parviflorus*, etc.

All that remains to say about the Latin (or Romanized Greek) noun concerns terms used unchanged in the nominative singular and nominative plural case forms, e.g.:

1. *Like RADIUS–RADII*
   alveolus, calculus, homunculus, locus, stimulus.
2. *Like LARVA–LARVAE*
   ala, antenna, axilla, bulla, bursa, formula, gastrula, macula, mamma, nebula.
   Note (i) *area-s* has become anglicized and many Americans use *formulas* in preference to *formulae*.
3. *Like DATUM–DATA*
   antrum, labium, maximum, medium, minimum, ostium, rostrum, septum, spectrum, stratum, cilium.
   Note (ii) *agenda* (like *data*) is a plural like *scissors* and *trousers*, whence *these agenda, these data, these strata*, NOT *this agenda, this data, this strata*.
4. *Like APEX–APICES*
   appendix, codex, index, matrix.
5. *Like AXIS–AXES*
   Avis thesis.

With the end we have here in view, there is little need for much detailed reference to the immensely luxuriant[1] flexional system of the Greek and the Latin verb. In German, French and Spanish dictionaries the flexional form listed is the one called the *infinitive*, i.e. the form which follows *to* in Anglo-American. Thus the dictionary entries *machen* (German), *faire* (French), *hacer* (Spanish) mean *to make*. If a verb is regular, i.e. conforms to one or other standard pattern (*conjugation*) the number following indicates which. If it does not do so, the dictionary cites sufficient examples of its vagaries as the *Concise Oxford Dictionary* cites *gave* and *given* after *give*.

In Latin and Greek dictionaries the main entry is the first-person singular of the present tense, e.g. Latin *facio* and Greek ποιεω (= *I make*). When a Latin verb is irregular, the flexional form usually most close to its contribution to the build-up of a technical term among the secondary entries is the one called the *supine* and listed last. Thus we read after *facio*: *facere* (infinitive), *feci* (past perfect) and *factum*. The supine *factum* has the stem we recognize in *artefact* or *factory*. The dictionary form of a Greek verb is commonly an adequate guide to incorporation in technical terms. If we speak of *give–gave–given* as the *principal parts* of the English verb *give*, the following illustrates what Greek grammarians cite as the principle parts (*Active* voice):

κλεπτω	(*I steal*);	κλειψω	(*I shall steal*),
ἐκλειψα	(*I stole*);	κεκλοφα	(*I have stolen*).

Only the first is formative, as in *kleptomania*.

Two suffixes added to the stem of a Latin word are worthy of mention because they are meaningful and as such occur in internationally current terms. A considerable class of Latin verbs referred to as *inceptive* carry the stem suffixes *-asc-, -esc-, -isc-*. These confer on the stem the notion of *beginning* or *becoming*. They go over to the present participle as *-escent*, etc., which we recognize in such adjectives as *putrescent, adolescent, iridescent*. The inceptive meaning of Greek verbs with the corresponding suffixes -σκ, -ισκ is rarely as transparent as in the Latin example cited, or as in γηρασκω, *I grow old*.

Many Latin nouns have *diminutive* forms derived by addition to the

---

[1] The English verb (excluding *be*, which is of mixed origin) embraces at most five forms (e.g. *give, giving, gives, gave, given*), more usually four (e.g. *guess guessing, guesses, guessed*). The Latin verb embraces 107, excluding all but one case form of the participles, supine, gerund and gerundive. The number of different forms of the Greek verb is much greater.

stem of *-ula, -ola, -ulus, -olus, -ulum, -olum*, according as the parent noun is feminine, masculine or neuter. Examples are:

filia	daughter	*filiola*	little daughter
rivus	stream	*rivulus*	brook
scalprum	knife	*scalpellum*	scalpel

When the stem of the parent noun ends in *l, n* or *r*, the diminutive terminal contracts to *-lla, -llus, -llum*, as in

corona	crown	*corolla*	little crown

Some nouns attach the diminutive suffixes *-cula, -culus, -culum*, e.g.:

corpus	body	*corpusculum*	particle
spes	hope	*specula*	ray of hope

Whence we derive *animalcule* from *animal*. Zoologists will also recall *cuniculus* in *Lepus cuniculus*, and *musculus* in *Mus musculus*. *Cuniculus* is a vulgar Latin word, probably cognate with *coney*. Its diminutive form recalls German *Kaninchen* (cf. *Mädchen*). From *musculus* (little mouse) we get *muscle*, suggested by the shape of the contracted biceps. From *caput* (abl. *capite*) for head, we get *capitulum*, the compact inflorescence of *Jasione, Dipsaceae* and *Compositae*.

PART TWO

Basic English–Greek and English–Latin Vocabularies

In this section the reader can gain a nodding acquaintance with a basic vocabulary of Greek and Latin other than names for living creatures and their parts, as set forth in the next section. To provide scope for word-building, the plan is classificatory, based partly on parts of speech (e.g. adjectives or verbs) and partly on categories of nouns (e.g. household articles or colours). Greek and Latin equivalents of English words appear in parallel columns, Latin on the extreme right, English on the extreme left.[1]

The grammatical conventions adopted are as follows. Apart from those of which the nominative singular (e.g. RADIUS and SPECTRUM) has come over into English unchanged, the case form cited for Latin nouns is the ablative (or common ablative–dative), this being the only singular case form which has survived in Spanish and Italian. For both Latin and Greek adjectives in (18), the case form cited is, as in lexicons, the nominative singular masculine; but in (19) the case form cited is the ablative singular masculine. As regards Latin verbs, the first form cited is the first person singular of the present tense, and this is the only one cited if the stem corresponds to the root of internationally current terms. If the stem of the present tense is not formative, or if more than one stem is formative, the second entry is either the supine (see p. 56), which ends in -*um*, or the first person singular of the perfect tense, which ends in -*i*. If a Latin nominative has passed into English unchanged, it appears in capitals.

For Greek nouns the entries are nominative singular case forms as in lexicons, followed by the genitive (or merely by the stem itself) if the stem of the latter is different. For Greek verbs, the entry is the first person singular of the present tense.

As regards Romanization of the Greek script, the aim has been to keep close to conventions that British science inherited from mediaeval Latin. As explained in Chapter 4, these include *a* for α, *i* for ι, *e* for ε or η, *o* for ο or ω and *y* for υ, except when ε or ο precedes υ. For

---

[1] The English meanings given are the ones retained in international usage, and in some cases these differ from the most usual meaning of the word in Greek.

diphthongs, $\alpha\iota$ becomes æ, $o\iota$ becomes œ and $o\upsilon$ becomes u. In accordance with a long tradition c replaces κ and ph replaces ϕ, though in some European countries scientists use k and f for κ and ϕ. American medical scientists prefer e for more traditional $\alpha\iota$ or $o\iota$. Since *paederasty* does not mean foot-fetish, nor does *paediatrics* mean chiropody, the writer prefers to stick to the ancient ways. However, he concedes to less conventional transliteration in one respect. Because our concern in this book is to recognize *roots* rather than case forms, the writer has refrained from Latinizing the noun and adjectival terminals η, os, $o\nu$ as *a*, *us*, *um*.

1. PARTICLES AND PRONOUNS USED AS PREFIXES

ABOVE, OVER	$\acute{\upsilon}\pi\epsilon\rho$	1. super
	(hyper)	superstructure, superscribe
	hypertension, hypertrophy	2. supra
		suprarenal, supramundane
ACROSS, THROUGH	$\delta\iota\alpha$	trans
	(dia)	transparent, transmit
	diagram, diagonal	
AFTER	$\mu\epsilon\tau\alpha$	post
	(meta)	post-mortem, post-natal
	metamorphic, metatarsal	
AGAINST	$\grave{\alpha}\nu\tau\iota$	contra
	(anti)	contradict, contravention
	antibiotic, antipathy	
AROUND	$\pi\epsilon\rho\iota$	circum
	(peri)	circumference, circumscribe
	perihelion, peripatetic	
AWAY FROM	$\grave{\alpha}\pi o$	a *or* ab
	(apo)	aberrant
	apostate, apostrophe	
BACKWARDS	$\acute{o}\pi\iota\sigma\theta\epsilon\nu$	retro
	(opisthen)	retrograde, retrospect
	opisthograph, opisthocoelous	
BEFORE	$\pi\rho o$	1. ante
	(pro)	antecedent, antenatal
	prolepsis, prologue	2. præ (pre-)
		preposition, preface
BELOW, UNDER	$\acute{\upsilon}\pi o$	sub
	(hypo)	subterranean, submarine
	hypoblast, hypodermis	
BESIDE, NEAR	$\pi\alpha\rho\alpha$	proxime
	(para)	approximate, proximity
	parabola, paratyphoid	

BETWEEN		inter
		interchange, interact
BEYOND	ὑπερ	ultra
	(hyper)	ultramontane, ultra-violet
	hyperborean, hypertonic	
DOWN	κατα	de
	(cata)	descend, deposit
	catadromous, catalysis	
I	ἐγω	ego
	(ego)	
	egocentric, egoism	
IN	ἐν-, ἐμ-	in- or im-
	(en or em)	insert, impress
	endemic, empathy	
INSIDE	ἐνδον	intra
	(endon)	intramolecular, intravenous
	endoderm, endophyte	
ON	ἐπι	
	(epi)	
	epidermis, epiblast	
ON BEHALF OF, IN FAVOUR OF		pro
		pro-British
OUT OF	ἐξ, ἐκ	ex or e
	(ex or ec)	expand, elucidate
	exegesis, eccentric	
OUTSIDE	ἐκτος	extra
	(ectos)	extramural, extraordinary
	ectomorph, ectoplasm	
SELF	αὐτος	ipse
	(autos)	ipsilateral
	autostylic, autoerotic	
THROUGH	δια	per
	(dia)	pervade, persecute
	diagnosis, diarrhoea	
TOGETHER WITH	συν, συμ-	cum
	(syn or sym)	
	syndrome, sympathy	
UP	ἀνα	
	(ana)	
	anadromous, analysis	
WITHOUT	ἀ-, ἀν-	sine
	(a or an)	sinecure
	anhydrous, agonic	

2. NUMERALS

NUMBER	ἀριθμος (arithmos) *arithmetic*	numero numeral, innumerable
ONE	ἑν (hen) *henotheism, hendiadys*	unus unit, unitarian, unison
TWO	δυο, δι- (duo *or* di-) *dihedral, dioecious*	duo duet, duplicate
THREE	τρια (neut. of τρεις) (tria) *triad, trimerous*	tres trefoil, trellis
FOUR	τεσσαρες, τετρα- (tessares *or* tetra-) *tetrad, tetrahedron*	quattuor quatrain, quatercentenary
FIVE	πεντε (pente) *pentagon, pentadactyl*	quinque quinquennial, quinquifid
SIX	ἑξ (hex) *hexagon, Hexagynia, Hexapoda*	sex sextet, sextant
SEVEN	ἑπτα (hepta) *heptarchy, Heptanchus*	septem September, septate, septenary
EIGHT	ὀκτω (octo) *octagon, octopus*	octo octant, octave
NINE	ἐννεα (ennea) *ennead, enneandrous*	novem November, novena
TEN	δεκα (deca) *decalogue, Decapod*	decem December, decennium
TWELVE	δωδεκα (dodeca) *dodecagon, dodecahedron*	duodecim duodecimo, duodenum
HUNDRED	ἑκατον (hecaton) *hectogram, hectare, hecatomb*	centum century, percentage
THOUSAND	χιλιοι (chilioi) *chiliad, kilometre*	mille millennium, millipede

FIRST	πρωτος (protos) *protoplasm, prototype*	primo primogeniture, primary
SECOND	δευτερος (deuteros) *deuterogamy, Deuteronomy*	secundo secondary
THIRD	τριτος (tritos) *tritagonist, tritium*	tertio tertiary, tertian
SINGLE	μονος (monos) *monochrome, monobasic*	singuli singular
DOUBLE	διπλοος (diploos) *diploblastic, diplodocus*	
TREBLE	τριπλοος (triploos) *triple, triploblastic*	triplico triplicate
HALF	ἡμι– (hemi-) *hemisphere, hemiplegia*	semi- semicircle, semifinal
SIMPLE, SINGLE	1. ἁπλοος (haploos) *Haplosporidia, haploid* 2. λιτος (litos) *Litopterna*	simplice, simplicity, simplicidentate

3. SHAPES

ANGLE	γωνια (gonia) *polygon, diagonal, trigonometry, agonic*	angulo angle, angular
BASE	βασις (BASIS)	BASE (abl.)
CENTRE	κεντρον (centron) *centre, egocentric, heliocentric*	1. centro centrifuge 2. medio medium, medial

CIRCLE	κυκλος	circulo
	(cyclos)	circular, circulate
	cyclic, tricycle,	
	Cyclostomata, epicycle	
CUBE	κυβος	cubo
	(cubos)	
	cubic, cubical	
CURVE	καμπυλη	sinu
	(campyle)	sine, sinusoidal,
	campylotropous,	sinus, insinuate
	campylospermous	
CYLINDER	κυλινδρος	cylindro
	(cylindros)	
	cylindrical	
LINE	γραμμη	linea
	(gramme)	linear
	Hexagrammus	
POINT	ἀκμη	1. acumine
	(ACME)	acumen, acuminiferous
		2. cuspide
		bicuspid, cuspidate, cusp
		3. mucrone (nom. mucro)
		unomucronate,
		mucronule
PYRAMID	πυραμις, πυραμιδος	pyramide
	(pyramis, -idos)	
RADIUS	ἀκτις, -ινος	RADIO
	(actis, -inos)	radius, radioactive
	actinotherapy, actinic,	
	Actinotrichia	
RHOMBUS	ῥομβος	RHOMBUS
	(rhombos)	
RING	γυρος	an(n)ulo
	(gyros)	annulate, annular
	gyrate, gyrostat,	
	Gyrocotyle	
SPHERE	σφαιρα	1. sphaera
	(sphaera)	
	ionosphere,	2. globo
	idiosphaerotheca	globular, globulin
SPIRAL	ἑλιξ, -ικος	1. spira
	(HELIX, -icos)	Spirogyra, spirochaete
	helical, helicopter	
		2. COCHLEA

66

SQUARE	τετραγωνον (tetragonon) *tetragonal*	1. quadrato quadrate
		2. quadrangulo quadrangle

4. COLOURS

COLOUR	χρωμα (chroma) *panchromatic, chromosome*	COLOR colour, colorimeter
WHITE	λευκος (leucos) *leucocyte, leucaemia*	1. albo albino, albumen 2. candido incandescent, candidate
BLACK	μελας (melas) *Melanesia, melanosis*	1. nigro denigrate, Nigeria 2. atro atrabilious, atrorubent
RED	ἐρυθρος (erythros) *erythema, erythrocyte*	rubro rubric, ruby
REDDISH	πυρρος (pyrros) *pyrrhole, pyrrhotite*	RUFUS (nom.) rufous, ruficaudate
BLUE	κυανεος (cyaneos) *cyanosis, Cyanophyceae*	caeruleo cerulean
GREEN	χλωρος (chloros) *chlorophyll, chlorine*	1. viridi viridian 2. glauco glaucous, glaucoma
YELLOW	ξανθος (xanthos) *xanthophyll, xanthoderma*	luteo luteal, luteovirescent CORPORA LUTEA
GREY	1. πολιος (polios) *poliomyelitis, polioplasm* 2. φαιος (phaeos) *phaeophyll, Phaeophyceae*	cineraceo cineraria, cinereous
PURPLE	πορφυρεος (porphyreos) *porphyry, haematoporphyrin*	purpureo purpura

67

PINK	ῥοδοεις	roseo
	(rhodoeis)	roseola, roseochrome
	rhododendron, rhodopsin	
VIOLET	ἰοειδης	VIOLA
	(iodes)	
	iodine, iodopsin	

5. THE FOUR ELEMENTS AND RELATED WORDS

AIR	ἀηρ	AURA
	(aer)	
	aeroplane, aeronautics	aethere
		ether, ethereal
BREATH	1. πνευμα	1. HALITUS
	(pneuma)	halitosis
	pneumonia, pneumotaxis,	2. spiritu
	pneumatopyle	spirit, respiration,
	2. πνοη	spirograph, spirometer
	(pnoe)	3. ANIMA
	hyperpn(o)ea, apn(o)ea	
WIND	ἀνεμος	1. flatu
	(anemos)	afflatus, inflate
	anemone, anemophilous,	
	anemometer	2. vento
		ventilate, ventifact
EARTH	γη	1. terra
	(ge)	terrestrial, territorial,
	geology, geography,	Mediterranean, terrain
	apogee	
		2. tellure
		tellurion, tellurium
CAVE	σπηλαιον	1. ANTRUM (nom.)
	(spelaeon)	2. spelunca
	spelaeology, spelaean	speluncar
		3. caverna
		cavern
CHALK	γυψος	creta
	(gypsos)	cretaceous, cretify
	gypsum, Gypsophila	
CLAY	πηλος	argilla
	(pelos)	argillaceous, argillite
	pelolithic	
CLEFT	σχισμη	fisso
	(schisme)	fissure, fissiparous
	schism, schismatic	
		HIATUS

	χασμα	rima
	(chasma)	rimate, rimiform, rimose
	chasm, chasmophyte,	
	chasmogamy	
GROUND SOIL	1. χθων	HUMUS
	(chthon)	
	autochthonous, chthonic	
	chthonophagia	
	2. χωριον	
	(CHORION)	
	chorography, chorology	
ON THE GROUND	χαμαι	
	(chamae)	
	chamaephyte, chameleon	
MINE	μεταλλεια	metallo
	(metalleia)	
MUD	ιλυς	limo
	(ilys)	limicolous, limivorous
	Ilysia, Ilyophis	
PEBBLE	ψηφος	CALCULUS
	(psephos)	
	Psephurus	
ROCK	πετρα	1. LAPIS (*nom.*), lapidi (*abl.*)
	(petra)	lapidary
	petrify, petrology	2. saxo
		saxatile, saxicolous,
		Saxifraga
		3. rupe
		rupestrine, rupicolous
SAND	ψαμμος	1. sabulo
	(psammos)	sabuline, sabulose
	psammophilous,	
	psammophyte	
		2. ARENA
FIRE	πυρ	igni
	(pyr)	igneous, ignite
	pyrex, pyrometer	
ASHES	1. σποδος	cinere
	(spodos)	cineraria, cinereous
	spodogenic, spodumene,	
	spodium	
	2. τεφρα	
	(tephra)	
	tephrite, tephromancy	

FLAME	φλοξ, φλογος	flamma
	(PHLOX, phlogos)	inflammable, flammeous
	phlogiston, phlogosis	
HEAT	θερμοτης	CALOR (nom.)
	(thermotes)	calorie
	thermostat, thermometer	
SMOKE	καπνος	fumo
	(capnos)	fumes, fumigate
	capnomancy, capnomor	
SOOT	1. ἀσβολος	fuligine
	(asbolos)	fuliginous
	asbolite, asboline	
	2. λιγνυς	
	(lignys)	
	pyroligneous	
SPARK	σπινθηρ	SCINTILLA
	(spinther)	scintillate
	spinthariscope	
WATER	ὑδωρ, ὑδατος	AQUA
	(hydor, hydatos)	aquiferous, aquarium,
	hydrogen, dehydrate,	aqueduct
	anhydrous	
BUBBLE	φυσαλλις	BULLA
	(physallis)	bullate, bullescence
	Physalia, physaliphore	
DEW	δροσος	rore (nom. ros)
	(drosos)	rosemary, roriferous, roscid
	Drosera, Drosophila	
DROP	σταγμα	GUTTA
	(stagma)	guttiform, guttulate
FOAM	ἀφρος	spuma
	(aphros)	spume, spumescence
	aphrosiderite, Aphrodite	
FOUNTAIN,	πηγη	fonte
SPRING	(pege)	fonticulus, font
	Pegasus	
ICE	κρυσταλλος	glacie
	(crystallos)	glacier, glaciation
	crystalline	
RAIN	ὑετος	pluvia
	(hyetos)	pluvial, pluviometer
	isohyetal, hyetology	
STEAM, VAPOUR	ἀτμος	VAPOR
	(atmos)	evaporate
	atmosphere, atmometer	

70

WAVE	κυμα	1. unda
	(cyma)	undulation, undulant
	cymose, kymograph	
		2. fluctu
		fluctuate
WHIRL, EDDY	δινη	verticillo
	(dine)	verticillaster, verticillate

6. TIME

AUTUMN	ὀπωρα	auctumno
	(opora)	
DAWN	ἠως	aurora
	(eos)	auroral
	eocene, Eohippus	
DAY	ἡμερα	die
	(hemera)	diurnal, diary
	hemeralopia, ephemeral,	
	hemerocallis	
EVENING	ἑσπερα	vespere
	(hespera)	vespers, vespertine
	Hesperian, Hesperornis	
HOUR	ὡρα	hora
	(hora)	
	horology, horoscope	
MONTH	μην	mense (pl. MENSES)
	(men)	menstruation
	menolipsis, catamenia	
MORNING	ἠως	matutino (adj.)
	(eos)	matutinal
	Eocene, eohippus	
NIGHT	νυξ, νυκτος	nocte
	(nyx, nyctos)	noctilucent, nocturnal
	nyctanthous, Nycticebus	
NOON	μεσημβρια	meridie
	(mesembria)	meridian, ante meridiem (a.m.)
	mesembryanthemum	
SPRING	ἑαρ	verno (adj.)
	(ear)	vernal
SUMMER	θερος	aestate
	(theros)	aestival, aestivation
TIME	χρονος	tempore
	(chronos)	temporary
	chronology, chronometer	

TWILIGHT	κνεφας	crepusculo
	(cnephas)	crepuscular
WEEK	ἑβδομας	hebdomade
	(hebdomas)	
	hebdomadal	
WINTER	χειμων	hieme
	(cheimon)	hiemal
YEAR	ἐνιαυτος	anno
	(eniautos)	annual, superannuated
	ἐτος, ἐτεος	
	(etos, eteos)	
	etesian, eteostic	

7. WEATHER

EAST WIND	εὐρος	euro
	(euros)	
FROST	παγος	pruina
	(pagos)	pruinate, pruinose
	pagoplexia	
HAIL	χαλαζα	grandine
	(CHALAZA)	
	chalazion, chalazodermia	
LIGHTNING	ἀστραπη	fulmine
	(astrape)	fulminate, fulminic acid
MIST	ὁμιχλη	NEBULA
	(homichle)	
NORTH WIND	βορεας	aquilone
	(boreas)	
	boreal, hyperborean	
RAINBOW	ἱρις	iride
	(IRIS)	iridescent
SHOWER	ὀμβρος	imbre
	(ombros)	
	ombrology, ombrometer	
SNOW	χιων	nive
	(chion)	
	chionablepsia, chionophobia	
SOUTH WIND	νοτος	noto
	(notos)	
STORM	χειμων	tempestate
	(cheimon)	tempest, tempestuous
SUNBEAM, RAY	ἀκτις, ἀκτινος	
	(actis)	
	actinic, Actinozoa	

THUNDER	βροντη	tonitru
	(bronte)	
	brontosaurus, brontophobia	
THUNDERBOLT	κεραυνος	
	(ceraunos)	
	ceraunoscope, ceraunics	
WEST WIND	ζεφυρος	zephyrus
	(zephyros)	
	zephyr	
WHIRLWIND	λαιλαψ	1. TURBINE (abl.)
	(lailaps)	
		2. VORTEX, *abl.* vortice
		Vorticella

8. THE UNIVERSE

CLOUD	1. *νεφος*	1. NIMBUS
	(nephos)	2. nube
	nephoscope	obnubilate
	2. *νεφελη*	
	(nephele)	
	nephelometer	
ECLIPSE	ἐκλειψις	defectione
	(ecleipsis)	
	eclipse, Ecliptic	
HEAVEN	οὐρανος	caelo
	(ouranos)	celestial
	uranium, uranometry	
MOON	σεληνη	luna
	(selene)	lunar, lunatic
	selenium, selenography	
SHADOW	σκια	umbra
	(scia)	umbrage, penumbra
	sciagraphy, skiascope	
SKY	αἰθηρ	1. caelo
	(aether)	
	ether, ethereal	2. aethere
STAR	ἀστηρ	1. STELLA
	(aster)	stellate, stelliform
	astronomy, asterisk,	
	astrolabe	2. sidere
		sidereal, siderostat
SUN	ἡλιος	sole
	(helios)	solar, solarization
	heliotrope, heliograph,	
	helium	

SUNBEAM	ἀκτις, -ινος	radio solis
	(actis, -inos)	
	actinograph, actinium	
UNIVERSE, WORLD	κοσμος (cosmos) *cosmopolitan, cosmic*	mundo mundane

9. GEOGRAPHICAL NAMES

BAY	κολπος (colpos)	sinu
BOUNDARY	ὁρισμα (horizma) *horizon*	TERMINUS terminal, indeterminate
CAPE	ἀκρα (acra)	promontorio promontory
CLIFF	κρημνος (cremnos) *cremnophobia*	caute
CONTINENT	ἠπειρος (epeiros) *epeirogeny*	continens (terra) continent
CURRENT	ῥευμα (rheuma) *rheumatism*	AESTUS
FIELD	ἀγρος (agros) *agromania, agronomy*	1. agro agriculture 2. prato Geranium pratense, pratincole
FLOOD	κατακλυσμος (cataclysmos) *cataclysm*	DILUVIUM antediluvian
FOREST, WOOD	ὑλη (hyle) *hylophyte, Hyla*	1. silva silvan, silvatica 2. nemore Nemorosa, Nemophila
GROVE	ἀλσος (alsos)	luco
HILL	βουνος (bounos) *bunodont, bunoid*	1. colle colliculus, colline 2. TUMULUS 3. CLIVUS declivity

74

ISLAND	νησος (nesos) *Melanesia, Micronesia,* *Polynesia*	insula insular, insulation, insulin
ISTHMUS	ισθμος (isthmos)	isthmo
LAKE, POOL	λιμνη (limne) *limnoplankton, limnetic*	1. lacu lake, lacustrine 2. LACUNA
LAND, EARTH	γη (ge)	terra Mediterranean, terrestrial
MARSH	ἑλος (helos) *helophyte*	palude paludal, paludrine
MOUNTAIN	ὀρος (oros) *orology, orography*	monte monticule, Monte Carlo
OCEAN	ὠκεανος (oceanos)	oceano (adj.)
PENINSULA	χερσονησος (chersonesos) *Chersonese*	paeninsula
PIT, CAVITY	κοιλωμα (coeloma) *coelom*	1. FORAMEN Foraminifera 2. puteo
PLAIN	πεδιον (PEDION)	CAMPUS campestral
RIVER	ποταμος (potamos) *Hippopotamus, Potamogeton,* *Mesopotamia*	1. flumine 2. fluvio fluvial
ROAD	ὁδος (hodos) *cathode, hodograph*	1. VIA 2. itinere itinerary, itinerant
SEA	1. θαλασσα (thalassa) *Thalassema, thalassin* 2. πελαγος (pelagos) *pelagian, archipelago*	1. mare Weston-super-mare, marine 2. ponto pontic, rhaponticin
SHORE	1. αἰγιαλος (aegialos) 2. ἀκτη (acte)	litore littoral

STRAIT	στενον	freto
	(stenon)	fret
STREAM	ῥεος	fluento
	(rheos)	confluent
	rheostat, rheoscope	
VALLEY	ἀγκος	valle
	(angcos)	vallecula
WATERFALL	καταρακτης	cataracta
	(cataractes)	
	cataract	
WHIRLPOOL	δινη	1. VORTEX
	(dine)	2. gurgite
	dinetic	regurgitate
WOOD, *see*		
FOREST		

★   ★   ★

EAST	ἀνατολη	oriente
	(anatole)	oriental
	Anatolia	
NORTH	ἀρκτος	septentrione
	(arctos)	septentrional
	arctic	
SOUTH	νοτος	1. austro
	(notos)	Australia
	notornis, Notoryctes,	2. meridie
	notogaea	meridional
WEST	δυσις	occidente
	(dysis)	occidental

10. MATERIALS AND SUBSTANCES

AGATE	ἀχατης	
	(achates)	
ALABASTER	ἀλαβαστριτης	
	(alabastrites)	
ALUM	στυπτηρια	alumine
	(stypteria)	alum, aluminium
AMBER	ἠλεκτρον	electro
	(ELECTRON)	
	electric	
AMETHYST	ἀμεθυστος	amethysto
	(amethystos)	
ASBESTOS	ἀσβεστος	ASBESTOS (nom.)
	(ASBESTOS)	

LINEN	βυσσος (byssos) *byssus, byssal*	LINUM linen
COAL	ἀνθραξ (ANTHRAX) *anthracite*	carbone carbon, carboniferous
COPPER	χαλκος (chalcos) *chalcolithic, Chalcidoidea*	Cupro– cupriferous, cupreous
CRYSTAL	κρυσταλλος (crystallos)	crystallo
DIRT	1. βορβορος (borboros) 2. μυσος (mysos) *mysophobia*	SORDES sordid, sordor
DUST	κονις (conis) *coniospermous, coniomycetes*	pulvere pulverize, pulverescent
FAT	1. λιπος (lipos) *lipolysis, lipase* 2. στεαρ, στεατ- (stear, steat-) *stearin, steatorrhoea*	1. adipe adipose, adipescent 2. SEBUM sebaceous, seborrhoea
FIBRE	νευρον (neuron)	fibra *fibriform, fibrillate,* *fibrositis*
FLAX	λινον (linon) *linen*	LINUM linoleum
FLINT	πυριτης (PYRITES)	silice silica, silicon
GLASS	ὑαλος (hyalos) *hyaline, hyaloid*	vitro vitreous, vitriol
GLUE	κολλα (colla) *colloid, collenchyma*	glutine glutinous, agglutinate
GOLD	χρυσος (chrysos) *chrysanthemum, chrysalis*	auro auriferous, aureole
HAY	χορτος (chortos)	f(o)eno

IRON	σιδηρος (sideros) *siderophil, sideroscope*	FERRO- ferrous, ferriferous, ferronickel
LEAD	μολυβδος (molybdos) *molybdenum*	plumbo plumb, plumbiferous
LEATHER	χοριον (CHORION) *choroid*	
LIME	τιτανος (titanos)	CALX (nom.), calce (abl.) calcareous, calcium
MAGNET	λιθος μαγνητις (lithos magnetis) *magnetic*	
MARBLE	μαρμαρον (marmaron)	marmore marmoreal
MERCURY	ὑδραργυρος (hydrargyros)	HYDRARGYRUM (= Hg)
PEARL	μαργαριτης (margarites) *margarine, Margaret*	margarita
SALT	ἁλς, ἁλος (hals, halos) *halogen, halophyte*	SAL, *abl.* sale saline, sal volatile
SALTPETRE	νιτρον (nitron) *nitrogen, nitric*	nitro-
SILK	μεταξα (metaxa)	1. BOMBYX (nom.) bombazine, bombic acid 2. serico holosericeous, serific
SILVER	ἀργυρος (argyros) *argyrodite, argyranthous*	argento argentite, argentiferous
STONE	λιθος (lithos) *lithograph, Neolithic*	1. saxo saxifrage, saxicolous 2. LAPIS (lazuli) lapidary, lapillus 3. PUMICE (abl.) 4. CALCULUS
STRAW	καρφος (carphos) *carpholite, carphology*	culmu culm

78

SULPHUR	θειον (theion) *thiourea, thiosulphuric*	sulfure
TIN	κασσιτερος (cassiteros) *cassiterite*	stanno stannic, stannite
VERMILION	κινναβαρι (cinnabari) *cinnabar*	
WAX	κηρος (ceros) *cerotic acid, Ceroxylon*	cera cerate, ceraceous
WOOD, TIMBER	1. ξυλον (xylon) *xylem, xylophone* 2. ὑλη (hyle) *hylophagous, hylotomous*	ligno lignin, lignite, ligneous
WOOL	1. ἐριον (erion) *Eriocaulon, Eriophyes* 2. μαλλος (mallos) *mallophagous*	lana lanolin, lanate

11. FOOD AND DRINK

BREAD	ἀρτος (artos) *artophagous, artocarpad*	pane pantry, pannier, marzipan
BROTH	ζωμος (zomos)	ius (nom.) juice
BUTTER	βουτυρον (butyron) *butyrinase, butyric*	butyro
CAKE	μαζα (maza)	PLACENTA
CHEESE	τυρος (tyros) *tyrosine, tyramine*	caseo caseous, casein
CREAM		spuma spume
DRINK	ποσις (posis)	potione potion

79

FEAST	εὐωχια	convivio
	(euochia)	convivial
FLOUR	παλη	1. farina
	(pale)	farinaceous, farinose
		2. POLLEN
FOOD	1. τροφη	1. esca
	(trophe)	esculent
	trophoblast, hypertrophy	
	2. σιτος	2. cibo
	(sitos)	cibation, cibarious
	parasite, sitophobia	
	3. ὀψον	3. alimento
	(opson)	alimentary
	oligopsony, opsonin	
HONEY	μελι	melle
	(meli)	mellifluous, melliferous
	melilot, meliphagous	
HUNGER	λιμος	fame
	(limos)	famished, famine
	limophthisis	
JUICE	1. χυλος	suc(c)o
	(chylos)	succulent, succiferrous
	chyle, chylopoietic	
	2. χυμος	
	(chymos)	
	chyme, chymotrypsin	
MEAL	1. ἀμυλος	
	(amylos)	
	amylaceous, amylase	
	2. ἀλφιτον	farina
	(alphiton)	
	alphitomorphous,	
	alphitomancy	
MEAT, FLESH	κρεας	carne
	(creas)	carnivorous, carnation
	creatine	
MILK	γαλα	lacte
	(gala)	lactic, lactose
	galaxy, galactose	
MUSTARD	σιναπι	sinape
	(sinapi)	
	sinapism, sinapisine	
OIL	ἐλαιον	oleo
	(elaeon)	oleaginous, carbolic, linoleum
	elaioplast, elaeoblast	

PEPPER	πεπερι (peperi)	pipere piperaceous
SPICE, SEASONING	ἀρωμα (AROMA) aromatic	1. AROMA 2. condimento condiment
SUGAR	σακχαρ (sacchar) saccharine, polysaccharide	saccharo
THIRST	διψα (dipsa) dipsomania, dipsopathy	siti
VINEGAR, ACID	ὀξος (oxos) oxalic, Oxalis	aceto acetic, acetose
WINE	οἰνος (oenos) oenology (or oinology) oenomania, oenophilist	vino vine, viniculture
YEAST, LEAVEN	ζυμη (zyme) enzyme, zymogenic	fermento fermentation

12. HOUSEHOLD EQUIPMENT AND CLOTHES

(A)

BAG	1. μαρσιπος (marsipos) marsupial, marsupium 2. ἀσκος (askos) Ascomycetes, ascocarp	1. sacco saccule, saccate 2. folliculo follicle 3. utre utricle, utriform
BASIN, BASKET	1. κοφινος (cophinos) coffin 2. καλαθος (CALATHOS) calathiform	1. PELVIS (nom.) 2. corbe corbula, corbicula 3. canistro canister
BED	κλινη (cline) clinic, diclinous	cubile
BLANKET, COVERING		STRATUM (nom.) stratified

Box	1. ἀγγειον	1. cista
	(angeion)	Cistudo
	angiocarpous, angiosperm	2. capsa
	2. θηκη	capsule, incapsulate
	(thece)	3. arca
	endothecium, theca	arcanum, arcane
	3. πυξις	
	(pyxis)	
	pyxidium, pyxis, pyxinia	
	4. κιστη	
	(kiste)	
	cistern, cist	
BROOM		scopae (pl.)
		scopiform, scopuliferous
CANDLE or	λυχνος	candela
LAMP	(lychnos)	candelabrum
	Lychnis, lychnoscope	
CARPET	ταπης	1. stragulo
	(tapes)	2. tapete
	tapestry	tapetum
CHAIR	καθεδρα	SELLA
	(cathedra)	
	cathedral, ex cathedra	
CUP	1. σκυφος	calice (nom. calix)
	(scyphos)	caliciform, calyx
	scyphiform, scyphistoma	
	2. κοτυλη	
	(cotyle)	
	cotyledon, cotyloid	
CURTAIN		VELUM
		veliferous, veligerous
DISH	πιναξ, πινακος	1. PATELLA
	(pinax, pinacos)	2. PATINA
	pinacocytes, pinacotheca	
FORK	δικελλα	1. furca
	(dicella)	furcula, bifurcate
		2. tridente
		trident
IMAGE	1. εἰδωλον	1. imagine
	(idolon)	image, imaginary
	idol, idolatrous	2. effigie
	2. εἰκων	effigy
	(icon)	3. SIMULACRUM
	icon, iconoclast	

82

KNIFE	μαχαιρα (machaera)	1. cultro coulter, cultrate 2. scalpro scalpel
LAMP	1. λαμπας (lampas) *lamp* 2. λυχνος (lychnos) *lychnidiate*	lucerna
MIRROR	ἐσοπτρον (esoptron)	SPECULUM
PILLOW, CUSHION	προσκεφαλαιον (proscephalaeon)	1. PULVINUS 2. PULVILLUS
SPOON	τορινη (torine)	cochleare Cochlearia
TABLE	τραπεζα (trapeza) *trapezium, trapeze*	1. tabula tabulate 2. mensa commensal
VESSEL OR BOWL	1. κρατηρ (CRATER) 2. κυτος (cytos) *cytology, leucocyte*	VAS, *pl.* VASA vascular

(B)

APPAREL, CLOTHES	ἐσθησις (esthesis)	veste divest, vestments
ARMOUR	πανοπλια (panoplia) *panoply*	armatura armature
BELT, GIRDLE	1. μιτρα (mitra) 2. ζωνη (ZONE) *zonociliate, zonule*	1. CINGULUM 2. balteo belt 3. cincto cincture
BOOT	κοθορνος (cothornos)	1. caliga 2. OCREA ocreate
CAP	πιλος (pilos) *Pilobolus, Pilocarpus*	PILEUS/PILEUM pileate, depilatory

CLOAK	χλαμυς (Chlamys) *chlamydospore, Chlamydera*	PALLIUM palliopedal, palliobranchiate
COVERING	καλυμμα (calymma) *kalymmocyte*	1. TEGMEN 2. VELAMEN 3. INVOLUCRUM involucellate
CROWN	στεμμα (stemma)	diademate diadem
GLOVE	χειροκτιον (chiroction)	
HELMET	1. κορυς (corys) *corystoid* 2. πηληξ (pelex) *Peleseia*	GALEA galeate
RING, SIGNET	δακτυλιος (dactylios) *dactylioglyph, dactyliography*	ANNULUS annular, Annuloida
ROBE	1. πεπλος (peplos) 2. στολη (STOLE)	
SEAM	ῥαφη (RAPHE) *araphorostic*	
SLIPPER, SHOE	ὑποδημα (hypodema)	1. calceo calceolaria, calceiform 2. socco sock
TASSEL	θυσανος (thysanos) *Thysanura, Thysanoptera*	
TUNIC	χιτων (CHITON) *chitin, Rhodochiton*	TUNICA tunic, Tunicata
VEIL	καλυπτρα (CALYPTRA) *calyptrogen, calyptriform*	integumento integument
WEB	ἱστος (histos) *histogenic, histolysis*	1. textura texture 2. TELA telarian

WREATH	στεφανος	1. TORQUE (abl.)
	(stephanos)	torquate
	Stephanotis, Stephanolepis	2. CORONA
		coronary, coronation

13. BUILDINGS AND THEIR PARTS

(A)

BARN	ἀποθηκη	granariis (abl. pl.)
	(apothece)	granary
	apothecium, apothecary	
BRIDGE	γεφυρα	PONS, (abl.), ponte
	(gephyra)	pontic, PONS VAROLII
	gephyrocercal, gephyrean	
BUILDING	οἰκοδομημα	aedificio
	(ecodomema)	edifice
CHURCH	ἐκκλησια	1. ecclesia
	(ecclesia)	2. BASILICA
	ecclesiastical	
GRAVE	ταφος	sepulcro
	(taphos)	sepulcre
	cenotaph, epitaph	
HOUSE	οἰκος	1. domo
	(oecos)	domestic, domicile
	economy, monoecious,	2. VILLA
	androecium	
PALACE	βασιλικη	1. regia
	(basilike)	2. BASILICA
PRISON	εἰρκτη	carcere
	(heircte)	incarcerate
SANCTUARY	ἀσυλον	ASYLUM
	(asylon)	
SHOP		officina
		Asparagus officinalis
TAVERN, INN	καταλυμα	hospitio
	(catalyma)	hospitable, hospital
TEMPLE	ναος	1. templo
	(NAOS)	temple
		2. fano
		fane
THEATRE	θεατρον	1. theatro
	(theatron)	2. ARENA
		3. spectaculo
		spectacular

85

(B)

ALTAR	βαμος (bomos)	altare altar
ARCH	καμαρα (camara)	arco arc, arcuate
BEAM	δοκος (docos) *Docoglossa*	1. trabe trabecula, trabeated 2. transtro transom
BEDCHAMBER	θαλαμος (thalamos) *thalamus, thalamifloral*	cubiculo cubicle
BRICK	πλινθος (plinthos) *plinth*	
CISTERN	δεξαμενη (dexamene)	cisterna
COLUMN, PILLAR	στηλη (STELE)	1. columna 2. pila pilaster, pillar
DITCH	ταφρος (taphros)	FOSSA fosse
DOOR	θυρα (thyra) *thyroid, thyridium*	1. OSTIUM ostiary, ostiole 2. porta portal
FLOOR	ἐδαφος (edaphos) *edaphology, edaphic*	pavimento pavement
GARDEN	κηπος (cepos)	hortu horticulture, hortensial
GATE	1. πυλη (pyle) *pylon, pylorus* 2. πυλωμα (pyloma) *pylome*	
HALL	αυλη (aule) *aularian, aulic*	ATRIUM
HEDGE, FENCE	φραγμα (PHRAGMA) *phragmoplast, diaphragm, Phragmatobia*	SEPTUM transept

OVEN	κλιβανος (klibanos)	furno furnace, furnage
PATH	ὁδος (hodos) *cathode, hodograph*	1. SEMITA 2. tramite
PICTURE		1. TABULA tabulate 2. pictura picture
PORCH	στοα (STOA) *stoic*	vestibulo vestibule
ROOF	στεγη (stege) *Stegocephali, Stegosaurus*	tecto tectrix, tectiform
ROOM	δωμα (doma)	1. conclavi conclave 2. cubiculo cubicle 3. dormitorio dormitory
STAIRCASE	κλιμαξ (CLIMAX)	scala scale, scalariform
STATUE	εἰκων (ICON) *iconoclasm, iconomania*	statua statue
STEP	κλιμακτηρ (climacter) *climacteric*	gradu gradually, centigrade
STUDY	μουσειον (museion)	1. BIBLIOTHECA 2. MUSEUM
THRESHOLD	οὐδος (udos)	LIMEN subliminal, liminal
TILE	κεραμος (ceramos) *ceramics*	1. IMBREX imbricate 2. TEGULA
WALL	τειχος (teichos) *teichopsia*	1. muro mural, immure 2. pariete parietal
WINDOW	σαλαμβη (salambe)	fenestra fenestration, Fenestella

ANCHOR	ἀγκυρα (ancyra) *ancyroid*	ancora
ARROW	ἰος (ios)	1. sagitta sagittate, sagittocyst 2. spiculo spiculiferous, spiculiform
AWL		subula subulate, subuliform
AXE	1. πελεκυς (pelecys) *pelecoid, Pelecypod* 2. ἀξινη (axine) *axiniform, axinomancy*	1. secure, -i securiform, Securifera 2. dolabra dolabriform
AXLE	ἀξων (AXON) *axoneme, axonost*	AXIS axipetal, axilemma
BAR, BOLT		claustra (n. pl.) claustrophobia
BELLOWS	φυσα (PHYSA) *Physalia, physeter*	folle follicle
BOW	τοξον (toxon) *toxaspire, toxophilite*	arcu arc, arch, arcuate
BREASTPLATE	θωραξ, θωρακος (THORAX, thoracos) *thoracic*	LORICA loricate
BRIDLE	ἡνια (henia)	1. FRENUM frenate, frenulum 2. LORUM lore, loral
CART	ἁμαξα (hamaxa)	plaustro
CHAIN	δεσμος (desmos) *desmognathous, desmid*	1. VINCULUM 2. CATENA catenary, concatenation
CHARIOT	ἁρμα (harma)	curru curricle, curriculum
CHISEL	γλυφειον (glypheion) *glyph, hieroglyphic*	SCALPRUM scalpriform

CLUB	1. ῥοπαλον (rhopalon) *Rhopalocera*	
	2. κορυνη (coryne) *Corynebacterium, Syncoryne*	clava claviform, clavicorn
COMB	κτεις, κτεν- (cteis, cten-) *ctenoid, Ctenophora, ctenidium*	1. pectine pectinate, pectinirhomb 2. STRIGILIS
CORD		fune funicle, funicular
DEVICE	μηχανημα (mechanema) *mechanism, machine*	1. artificio artificial 2. instrumento instrument
DRUM	τυμπανον (tympanon) *tympanohyal, tympanectomy*	TYMPANUM
FILE	ῥινη (rhine)	lima limation, limail
FLUTE, (PIPE)	αὐλος (aulos) *aulostomatous, aulete*	TIBIA
FUNNEL	χοανη (choane) *choanocyte, Choanoflagellata*	infundibulo infundibuliform
GIMLET	τρυπανον (trypanon) *Trypanosoma, Trypanourgus*	
GOAD, STING	κεντρον (centron) *Centrophorus*	STIMULUS stimulate
HAMMER	σφυρα (sphyra) *Sphyraena*	MALLEUS malleoramate, malleolar
HINGE	στροφευς (stropheus)	CARDO, *abl.* cardine cardinal
HOOK	1. ἀγκιστρον (angcistron)	1. hamo hamate, hamirostrate
	2. ἁρπαγη (harpage) *harpagones, harpagon*	2. UNCUS uncinate, unciform

89

KEEL	τροπις (tropis)	CARINA cariniform, carinate
KEY	1. κλεις, κλειδος (cleis, cleidos) *cleido-mastoid, cleidoic* 2. κλειθρον (cleithron) *cleithrum, cleithral*	clave, -i clavichord, clavicle
KNOT	ἁψις (hapsis)	nodo node, nodule
LEVER	μοχλος (mochlos)	VECTIS
LID		OPERCULUM
LOOM	ἰστος (histos)*	(textorium instrumentum)
LUTE	κιθαρα (kithara) *guitar, zither*	cithara
LYRE	λυρα (LYRA) *lyric, lyrate, lyriform*	lyra
MACHINE	μηχανη (mechane) *mechanical*	1. machina machine 2. organo organ
MAST	ἰστος (histos)	malo
MISSILE	βελος (belos) *Belostoma, belomancy*	MISSILE
NEEDLE	ῥαφις, ῥαφιδ- (RHAPHIS) *raphidiferous,* *Rhaphidopus*	acu acute, acupuncture
NET	δικτυον (dictyon) *sarcodictium, dictyosome*	RETE retecious, reticulum
OAR	κωπη (cope) *Copepoda*	remo remiped, remex
PEG, STAKE	πασσαλος (passalos) *Passaloeeus, Passalidae*	cultello

* Also *web* or *warp*, whence *histology*.

PEN, STYLUS	γραφις, γραφιδος (graphis, graphidos)	1. penna (= quill) pen 2. STILUS (nom.) stylus, stylograph
PIN	περονη (PERONE) peroneal, Peronospora	CLAVUS
PLANE		runcina runcinate
PLOUGH	ἀρατρον (aratron)	1. aratro aration 2. vomere ethmovomerine, vomeronasal
PULLEY	τροχιλια (trochilia)	TROCHLEA
QUIVER	φαρετρα (pharetra)	pharetra
ROD	ῥαβδος (rhabdos) rhabdite, Rhabdocoela	1. virga virgate, virgula 2. FERULA ferule, ferulaceous
RUDDER	πηδαλιον (pedalion)	GUBERNACULUM
SADDLE		SELLA
SAIL	ἱστιον (histion)	velo velar, velamen
SAW	πριων (prion) priodont	SERRA serriform, serriferous
SHEATH	κολεος (coleos) Coleus, Coleoptera	VAGINA vaginipennate, vaginicolous
SHIELD	θυρεος (thyreos) thyroid	1. SCUTUM scutigerous, scutiped 2. UMBO umbonate
SHIP, SKIFF	1. σκαφη (scaphe) bathyscaphe, scaphoid 2. ναυς (naus) nausea	nave, -i navy, navigate
SHUTTLE	κερκις (cercis)	radio

91

SICKLE, SCYTHE	1. δρεπανον (drepanon) *drepanium, Drepanodon*	FALX (abl. falce) falciform, falcate
	2. ἁρπη (harpe) *harpes*	
SIEVE	1. κοσκινον (coscinon) *coscinomancy, Coscinodon*	cribro cribriform, cribellum
	2. ἠθμος (ethmos) *ethmose, ethmophract*	
SIPHON	σιφων (SIPHON) *siphonoglyph, siphonostomatous*	siphone
SLING, CATAPULT	σφενδονη (SPHENDONE)	catapulta
SPADE	σμιννη (sminye)	pala
SPEAR	δορυ (dory) *dorylaner*	1. hasta hastate, hastifoliate 2. lancea lanceolate, lance
SWORD, BLADE	1. ξιφος (xiphos) *Xiphosura, xiphoid, Xiphocercus* 2. σπαθη (SPATHE) *spathaceous*	1. GLADIUS, GLADIOLUS gladiator 2. spatha 3. ense ensiform 4. mucrone (nom. mucro) mucronate
TABLET, PLATE	πλαξ, πλακος (plax, placos) *placoganoid, placoderm*	lapide
THREAD	1. μιτος (mitos) *mitochondria, mitosis* 2. νημα, νηματος (nema, nematos) *chromonema, Nematoda*	1. filo filament, Filaria, filaceous 2. STAMEN 3. capillo capillary, capillose
TOOL	ὁργανον (organon) *organ, organism*	instrumento instrumental
TRAP, SNARE	παγη (page)	1. pedica 2. plaga

Trumpet	σαλπιγξ, σαλπιγγος (Salpinx, salpingos) *salpingostomy, salpiglossis*	Tuba
Tube, Pipe	σωλην (solen) *solenia, solenocyte,* *solenoid*	1. tubulo tubular, tubule 2. Fistula 3. canale canal, canaliform 4. tubo tubicorn, tubiform
Vehicle	ὀχημα (ochema)	vehiculo vehicle
Vice		Forceps
Weapon	ὁπλον (hoplon) *hoplognathous, hoplite*	telo
Wedge	1. σφην (sphen) *zygosphene, sphenoid,* *Sphenodon* 2. ἐμβολος (embolos) *embolism*	cuneo cuneiform, cuneate
Wheel	τροχος (trochos) *trochophore, trochoblast*	rota rotation, rotary
Whip	μαστιξ, μαστιγος (mastix, mastigos) *heteromastigate,* *mastigobranchia,* *Mastigophora*	Flagellum flagellant, Flagellata
Whistle	συριγξ, συριγγος (Syrinx, syringos) *syringe, syringium*	Fistula
Yoke	ζυγον (zygon) *zygote, zygapophysis,* *homozygous*	Iugum jugate, conjugate

15. PEOPLE AND SOCIETY

Actor	ὑποριτης (hypocrites) *hypocrite, hypocritical*	1. Actor 2. histrione histrionic

ALLY	συμμαχος	socio
	(symmachos)	sociable, associate
AMBASSADOR	πρεσβυς	legato
	(presbys)	legate, delegate
	presbyter	
APPARITION	φαντασμα	SIMULACRUM
	(phantasma)	
	phantasm, phantom	
ARTISAN	τεχνιτης	fabro
	(technites)	prefabrication, fabrile, fabric
	technician	
ATTENDANT	διακονος	1. Satellite
	(diaconos)	2. MINISTER
	diaconal, deacon	
BATTLE	συμβολη	pugna
	(symbole)	pugnacious
BATTLE ARRAY	ταξις	acie
	(TAXIS)	
	taxonomy, parataxis	
BISHOP,	επισκοπος	1. pontifice
OVERSEER	(episcopos)	pontificate, pontiff
	episcopal	2. episcopo
		episcopal
BRIDE	νυμφη	nupta
	(nymphe)	nuptial
	nymphomania, nymph	
BROTHER	αδελφος	fratre
	(adelphos)	fraternal, fraternity
	Philadelphia,	
	monadelphous	
BUILDER	τεκτων	1. aedificatore
	(tecton)	edificatory
	tectonic, architect	2. structore
		constructor
CHILD, BOY	παις, παιδος	1. infante
	(paes, paedos)	infant
	paedogenesis, paediatric	2. puero
		puerperal, puerile
CITIZEN	πολιτης	civi
	(polites)	civic, civilian
	politics, politician	
CITY	πολις	urbe
	(polis)	urban, conurbation
	metropolis	

94

COIN	δραχμη (drachme) *drachma)*	1. numisma numismatics
		2. nummo nummulitic, nummular
		3. stipe stipend, stipendiary
COMMANDER	στρατηγος (strategos) *strategy*	1. praefecto prefect
		2. imperatore imperative
		3. DUCE duke
COMMERCE	ἐμπορια (emporia) *emporium, emporetic*	commercio commercial
CUSTOM	νομος (nomos) *antinomy, antinomian*	1. instituto institution
		2. MORES (pl.) moral
		3. consuetudine consuetude
DEBT	χρεος (chreos)	1. alieno alienate
		2. debito debit, debt
DESPOT	δεσποτης (despotes) *despotic*	domino dominate, domineering, anno domini
DEVIL	διαβολος (diabolos) *diabolical, diabolism*	diabolo
DICTATOR	τυραννος (tyrannos) *tyrant, tyrannosaurus*	DICTATOR
ENEMY	πολεμιοι (pl. adj.) (polemioi) *polemical*	1. hoste hostile
		2. inimico inimical, enemy
FARMER	γεωργος (georgos) *George, georgics*	1. agricola
		2. colono colony, colonial
FATHER	πατηρ (pater)	1. patre patrimony, paternoster
		2. parente parent

FUNERAL	ἐκφορα (ecphora)	funere funeral, funerary
GIFT	δωρον (doron)	1. dono donation 2. munere remuneration
GUARD	φυλακτηρ (phylacter) *phylactocarp,* *phylactolaematous*	
GUEST	ξενος (xenos) *xenia*	
HERDSMAN	βουκολος (bucolos) *bucolic*	PASTOR
HOST	ξενος (xenos)	hospite hospitable, hospitality
HUMAN BEING	ἀνθρωπος (anthropos) *anthropology,* *Pithecanthropus*	HOMO, homine homicide, hominiform
INTERPRETER	προφητης (prophetes) *prophet*	interprete interpreter
JUDGE	κριτης (crites) *critic*	iudice judicial, sub-judice
KILLER	αὐτοχειρ (autocheir)	homicida homicide
KING	βασιλευς (basileus)	rege regal, regicide
LAW	νομος (nomos) *economy, autonomy*	1. iure jurisdiction, jury 2. lege legal
MAGICIAN	μαγος (magos) *magic*	MAGUS/MAGI
MALE, MAN	ἀνηρ, ἀνδρος (aner, andros) *androecium, androspore*	viro virile
MAN	ἀνθρωπος (anthropos) *anthropology, anthropoid*	humano humanity

96

MARKET	ἀγορα (agora) *agoraphobia*	FORUM
MASTER	κυριος (kyrios) *kyrie eleison*	1. domino dominion, domineer 2. magistro, magistrate
MESSENGER	ἀγγελος (angelos) *angel, evangelist*	nuntio nuncio, announce
MONEY, WEALTH	χρηματα (chremata) *chrematist, chrematistic*	1. pecunia pecuniary 2. moneta monetary
MOTHER	μητηρ (meter)	matre maternal, matron, pia mater
OLD MAN	1. γερων, γεροντ- (geron, geront-) *geriatrics, gerontology* 2. πρεσβυς or πρεσβυτης (presbys or presbytes) *presbyter, presbyopia*	sene senile, senescence
OVERSEER	ἐπισκοπος (episcopos) *episcopal*	1. CURATOR 2. praeside president
OWNER	κεκτημενος (cectemenos)	POSSESSOR
PARENT	γονευς (goneus)	parente parental
PAUPER	πενης (penes) *leukopenia*	
PEOPLE	δημος (demos) *demagogue, democracy*	populo popular, population
PHYSICIAN	ἰατρος (iatros) *paediatrician, psychiatric*	medico
PILOT	κυβερνητης (cybernetes) *cybernetics*	gubernatore governor
PRICE, COST	τιμη (time) *timocracy*	pretio depreciate, appreciation

PRIEST	ἱερευς	sacerdote
	(hiereus)	sacerdotal
	hierarchy, hieroglyphics	pontifice
		pontifical
PRISONER	δεσμωτης	captivo
	(desmotes)	captivate, captive
PUNISHMENT		1. poena
		penalty, subpoena
		2. castigatione
		castigate
		3. vindicta
		vindictive
REWARD, PRIZE	1. μισθος	Pʀ(ᴀ)ᴇᴍɪᴜᴍ
	(misthos)	
	2. ἀθλον	
	(athlon)	
	athlete, pentathlon	
RULER	1. ἀρχων	1. ʀᴇᴄᴛᴏʀ
	(archon)	2. ᴍᴏᴅᴇʀᴀᴛᴏʀ
	matriarchy, monarch	
	2. δυναστης	
	(dynastes)	
	dynasty	
SAILOR	ναυτης	nauta
	(nautes)	nautical
	astronaut, Nautilus	
SCULPTOR	ἀγαλματοποιος	ꜱᴄᴜʟᴘᴛᴏʀ
	(agalmatopoios)	
	agalmatolite	
SERVANT	διακονος	1. ᴍɪɴɪꜱᴛᴇʀ
	(diaconos)	2. ᴀᴅᴍɪɴɪꜱᴛᴇʀ
	deacon, diaconal	
SISTER	ἀδελφη	sorore
	(adelphe)	sorority
SLAVE	δουλος	1. servo
	(doulos)	servant, servile
	dulocracy	2. mancipio
		emancipate, manciple
		3. ᴀɴᴄɪʟʟᴀ
		ancillary
SOLDIER	1. στρατιωτης	1. milite
	(stratiotes)	military
	2. ὁπλιτης	2. pugnatore
	(hoplites)	

SON	υἱος (huios)	filio filial, affiliation
STATE	πολις (polis) *politics*	1. respublica 2. regno 3. civitate 4. imperio imperial
STEERSMAN	κυβερνητης (cybernetes) *cybernetics*	RECTOR rectrix
STEWARD	οἰκονομος (eeconomos) *economy, economical*	1. PROCURATOR 2. ADMINISTRATOR
STRANGER, FOREIGNER	1. ξενος (xenos) *xenophobia, xenolith* 2. βαρβαρος (adj.) (barbaros) *barbarian, barbarous*	1. hospite hospital, hospitable 2. peregrino peregrine, peregrinate
TEACHER	παιδαγωγος (paedagogos) *pedagogue*	1. DOCTOR 2. PROFESSOR 3. Praeceptor
THIEF	κλεπτης (cleptes) *cleptomania, clepsydra*	1. fure furtive, furuncle 2. clepta
TRIBE, CLAN	1. ἐθνος (ethnos) *ethnography, ethnology* 2. φυλον *phylum, phylogenesis*	1. tribu tribal, tribe 2. GENUS 3. STIRPS stirp
VERDICT, JUDGEMENT	κρισις (CRISIS)	1. sententia sentence 2. arbitrio arbitrate, arbitrary
VICTORY	νικη (nike)	1. victoria victory, victorious 2. triumpho triumph, triumphant
VILLAGE	κωμη (come)	1. pago pagan 2. vico vicinity

VIRGIN	παρθενος (parthenos) *Parthenon, parthenogenesis*	VIRGO (abl. virgine)
WANDERER	πλανητης (planetes) *planet*	errone erroneous
WOMAN	γυνη (gyne) *gynaecology, gynoecium*	1. muliere muliebrity 2. femina feminine, effeminate

16. LEARNING AND ART

ART	τεχνη (techne) *technology, polytechnic*	arte
BOOK	βιβλιον (biblion) *bibliography, bibliophile*	libro library
DANCE	χορος (choros) *choreography, chorus*	saltatione saltatory
DEBATE, ARGUMENT	ἐρις (eris) *eristic*	controversia controversial
DISCOURSE	λογος (logos) *logic, dialogue*	1. oratione oration 2. sermone sermon
FABLE	μυθος (mythos) *myth, mythology*	fabula fabulous, fable
HYMN	ὑμνος (hymnos) *hymn, hymnal*	hymno
LEARNING	μαθημα (mathema) *mathematics*	eruditione erudition, erudite
LETTER (ABC)	γραμμα (gramma) *grammar, telegram*	littera literary, literal
LETTER	ἐπιστολη (epistole) *epistolary, epistle*	epistola

100

PAPER	παπυρος	charta
	(papyros)	chart
	papyrus	
POEM	ποιημα	poemate
	(poema)	
PREFACE	προοιμιον	praefatione
	(proemion)	preface
	proem	
RHETORIC	ρητορικη	rhetorica
	(rhetorike)	
ROW, VERSE	στιχος	versu
	(stichos)	verse
	stichomythia, distichous	
RULE, ROD	κανων	regula
	(canon)	regulate
	canonical	
SCHOOL	σχολη	schola
	(schole)	
SCIENCE	ἐπιστημη	scientia
	(episteme)	
	epistemology	
SYLLABLE	συλλαβη	syllaba
	(syllabe)	
TRAINING	ἀγωγη	disciplina
	(agoge)	discipline
	pedagoguy	
UNDERSTANDING	φρην	mente
	(phren)	mental
	phrenology, phrenic,	
	oligophrenia	
WORD,	λογος	verbo
DISCOURSE	(logos)	verbal
	monologue, zoology	
WRITING	γραφη	scripto
	(graphe)	script, scripture
	autograph, biography	

17. ABSTRACTIONS

ACTION	πραξις	actione
	(praxis)	action
	apraxia, praxinoscope	
AGE (OLD)	γηρας	1. senio
	(geras)	senility
	geriatrics	2. senectute
		senectitude

AGREEMENT	ἁρμονια	CONSENSUS
	(harmonia)	
	harmony, philharmonic	
ARRANGEMENT	θεσις	dispositione
	(THESIS)	disposition
BATTLE	συμβολη	pugna
	(symbole)	pugnacious
BEGINNING	ἀρχη	1. inceptione
	(arche)	inception
	archenteron, archegonium	2. initio
		initial, initiate
BIRTH	1. γενετη	1. ortu
	(genete)	abortion
	2. τοκος	2. genere
	(tokos)	generate, regeneration
	arrenotokous, thelytokous	
BIRTHDAY	γενεθλη	
	(genethle)	
	genethliacal, genethlialogy	
BOND	δεσμος	copula
	(desmos)	copulate, copulative
	desmocyte, desmid	
BRIGHTNESS	γανος	CANDOR
	(ganos)	
	ganoid, placoganoid	
CARE,	θεραπεια	cura
ATTENDANCE	(therapeia)	cure, curate
	therapeutic, therapy	
CAUSE	αἰτια	causa
	(aetia)	cause
	aetiology	
CONDITION	ἑξις	STATUS
	(hexis)	
	cachexia	
CONTEST	ἀγων	contentione
	(agon)	contention, contend
	protagonist, agony	
DARKNESS	σκοτος	obscuritate
	(scotos)	obscurity
	scotophobia, scotometer	
DEATH	θανατος	1. morte
	(thanatos)	mortuary, postmortem
	thanatology, euthanasia	2. letho
		lethal
		3. nece

DIRECTION, TURNING	τροπη (trope) *heliotrope, phototropism*	
END, PURPOSE	τελος (telos) *teloblast, telophase*	fine final, finish, infinite
EXISTING THINGS	ὀντα (onta) *palaeontology, ontogenetic*	vita vital
FACT, DEED	πραγμα (pragma) *pragmatic, pragmatism*	facts factual
FEELING	αἰσθησις (aesthesis) *anaesthesia*	sensu sensual
FIGURE, IMAGE	πλασμα (PLASMA) *plasmolysis, plasmocyte*	1. IMAGO (abl. imagine) imaginary 2. effigie effigy
FORM, SHAPE	μορφη (morphe) *morphology, endomorph*	1. FACIES 2. forma formative 3. figura figurate
FORMATION	πλασις (plasis) *achondroplasia, hyperplasia*	conformatione conformation
FREEDOM, LIBERTY	ἐλευθερια (eleutheria) *eleutherodactyl, eleutherophyllous*	libertate liberty
GENERATION	γονη (gone) *gonad, gonangium*	
GIFT	δωρον (doron) *Dorothy*	1. dono donation 2. stipe stipendiary
GOD	θεος (theos) *theology, theosophy*	deo deism, deify
HATRED	μισος (misos) *misogynist, misanthropy*	ODIUM

HEAP		CUMULUS (nom.) accumulate, cumulonimbus
HEIGHT	ὕψος (hypsos) hypsography, hypsometer	1. FASTIGIUM fastigiate 2. culmine culminate auxiliary, auxin
HERD, FLOCK	1. ἀγελη (agele) 2. ποιμνη (poemne)	grege congregate, aggregate, gregarious
HOLE, CAVITY	1. τρημα (trema) Trematode, Monotreme 2. τρυμα yma) 3. κοιλωμα (coeloma) coelom	1. FORAMEN (pl. FORAMINA) 2. LACUNA
IMITATION	μιμησις (mimesis) mimetic, mimic, mime	imitatione imitation, imitate
KNOWLEDGE	1. ἐπιστημη (episteme) epistemic, epistemology 2. γνωσις (GNOSIS) diagnosis, prognosis	scientia science
LAMENT	ἐλεγος (elegos) elegy, elegiac	lamentatione lamentation
LAYER	ἐπιβολη (EPIBOLE) epibolic	1. LAMINA 2. CORIUM excoriate
LIFE	βιος (bios) biology, autobiography	vita vitality
LIGHT	φως, φωτος (phos, photos) phosphorus, photography	1. luce lucent, luciferase 2. lumen (nom.), lumine (abl.) luminiferous, illuminate
LONGING, DESIRE	ὀρεξις (orexis) anorexia	1. DESIDERIUM 2. LIBIDO libidinous

104

LOVE	ἔρως, ἔρωτος (EROS, erotos) *erotic, erogenous*	amore amorous
LUMP, CALLUS	τυλος (tylos) *Tylopoda*	1. CALLUS callosity 2. massa massive
MARK	στιγμα (STIGMA) *astigmatism*	
MARVEL	1. θαυμα, θαυματος (thauma, thaumatos) *thaumaturgy* 2. τερας (teras) *teratoma, teratology*	monstro monstrosity
MEMORY	μνημη (mneme) *mnemonic*	memoria memorial
MIND	ψυχη (PSYCHE) *psychiatry, psychosomatic*	1. ANIMUS 2. mente mental
MIXTURE	μιξις (mixis) *amphimixis, apomixis*	mixtura mixture
MODEL, TYPE	τυπος (typos) *typical, prototype*	EXEMPLUM exemplary
MUSIC	μουσικη (musice)	musica musical
NAME	ὀνομα, ὀνοματος (onoma, onomatos) *onomatopoeia, onomatic*	nomine nominate, nominal
NARRATIVE	ἱστορια (historia) *history, historian*	narratione narration, narrator
NATURE	φυσις (physis) *physics, physiognomy*	natura natural
NUMBER	ἀριθμος (arithmos) *arithmetic, arithmomancy*	numero numerals, supernumerary
ODOUR	ὀσμη (osme) *osmium, anosmia*	odore odoriferous

OFFSPRING, RACE	γενος (genos) *genocide, genotype*	progenie progeny, pregenitor
OMEN	τερας, τερατος (teras, teratos) *teratoid, teratoma*	OMEN ominous
OPINION	δοξα (doxa) *orthodox, paradox*	1. opinione opinion 2. sententia sententious, sentence
ORIGIN	γενεσις (GENESIS)	origine aboriginal
PAIN	ὀδυνη (odyne) *anodyne*	dolore dolorous
PART	μερος (meros) *meroblast, isomer*	parte particle, bipartite
PEACE	εἰρηνη (Irene) *(e)irenic, (e)irenarch*	PAX (abl. pace) pacify, pacifism
PERCEPTION	αἰσθησις (aesthesis) *anaesthesia, aesthete*	
PHRASE	φρασις (phrasis) *phraseology*	locutione circumlocution
PITY	ἐλεημοσυνη (eleemosyne) *eleemosynary*	misericordia misericord
PLACE	τοπος (topos) *topography, isotope*	LOCUS locomotive, location
POWER	δυναμις (dynamis) *thermodynamic, dynamo*	1. potestate 2. potentia potential, potentiometer
PROPOSITION	προβλημα (problema) *problem*	conditione conditional
PURIFICATION, CLEANSING	καθαρσις *catharsis*	1. purgatione purgative, purgatory 2. purificatione
QUANTITY	πληθος (plethos) *isopleth*	quantitate quantity

QUESTION		quaestione
RACE, RUNNING	δρομος (dromos) *hippodrome, dromedary* *anadromous, syndrome*	1. CURRICULUM curricle 2. cursu cursive, cursory
RANK, ARRANGE-MENT	ταξις (taxis) *taxonomy*	ordine ordinal
REFLECTION, CONTEM-PLATION	θεωρια (theoria) *theory, theorem*	1. cogitatione cogitation, cogitate 2. deliberatione deliberation 3. consideratione consider
RELEASE	λυσις (lysis) *dialysis, catalysis*	liberatione liberation
RHYTHM	ρυθμος (rhythmos)	rhythmo
RICHES	πλουτος (plutos) *plutocrat*	1. fortuna fortune 2. opulentia opulence
SECRET RITE	1. μυστηριον (mysterion) *mysteries* 2. ὀργια (pl.) (orgia) *orgy*	ritu ritual
SELF	αὐτος (autos) *automobile, autobiography*	sui suicide
SHAKING, SHOCK	1. σεισμος (seismos) *seismometer, seismic* 2. τρομος (tromos) *tromophonia, tromometer*	1. concussione concussion, concuss 2. succussione succussion, succussation
SIGN, SYMBOL	1. σημα (sema) *semantics, semaphore* 2. σημειον (semeion) *semeiology, semeiotics*	signo sign, signal

SIGNIFICATION	σημασια (semasia) semasiology	
SLEEP	1. ὑπνος (hypnos) hypnosis, hypnophobia	1. sopore soporific
	2. Μορφευς (Morpheus) morphia, morphine	2. somno somnolent, somnifacient
SONG	1. ψαλμος (psalmos) psalm, psalmody	cantu cantata, canticle, descant
	2. ᾠδη (ODE) melody, threnody	
SOUND	1. ἠχος (echos) echo, echolalia	1. sono sonorous, dissonant 2. strepitu
	2. φωνη (phone) euphonious, phonetic	strepitous, strepitant 3. STRIDOR strident, stridulation
SPACE	χωρος (choros) chorography, chorology	spatio spatial, spatiotemporal
SPEECH	1. φασις (phasis) aphasia, dysphasia	lingua bilingual
	2. λεξις (lexis) lexicon, alexia	
STRETCHING	τονος (tonos) tone, tonicity, auxotonic	intensione intense
SUFFERING	παθος (PATHOS) pathetic, pathology	toleratione toleration
TAX, TRIBUTE	1. φορος (phoros)	1. exactio exaction
	2. δασμος (dasmos)	2. tributo tribute 3. stipendio stipendiary
THING, OBJECT	χρημα (chrema)	re republic, real

THRUST	ὠσμος (osmos) *osmosis, osmometer*	
TOUCH	θιγμα (thigma) *thigmotropism, thigmaesthesia*	tactu tactile, contact
TRIAL	δικη (dice)	1. experientia experience 2. iudicio judiciary
TWIST	στροφη (strophe) *catastrophe*	
UNIT	μονας (monas)	1. monade monad 2. unione union
VIEW, SPECTACLE	ὁραμα (horama) *panorama, cyclorama*	spectaculo spectacle
VOICE		vox (nom.), voce (abl.) vocal, convocation
WAR	πολεμος (polemos) *polemic*	bello bellicose, belligerent
WEARINESS		taedium (nom.) tedium, tedious
WILL	θελημα (thelema)	voluntate voluntary, volunteer
WISDOM	σοφια (sophia) *philosophy, sophist*	sapientia sapient, Homo sapiens
WORK	ἐργον (ergon) *ergometer, ergophobia*	1. OPUS (*pl.* opera) operate 2. labore labour, elaborate
WORSHIP	λατρεια (latreia) *idolatry, heliolatry*	1. veneratione veneration 2. adoratione adoration

YOUTH	ἤβη	adolescentia
	(hebe)	adolescence
	hebetic, hebephrenia	

18. ADJECTIVES AND ADVERBS

ACID, SHARP	ὀξυς	1. acidus
	(oxys)	acid, acidophilic
	oxygen, Amphioxus	2. acerbus
		acerbity, exacerbate
AFAR (adv.)	τηλε	1. procul
	(tele)	2. longe
	telescope, telepathy	
AGAIN (adv.)	παλιν	iterum
	(palin)	reiterate
	palingenesis, palindrome	
ALIKE,	ὁμος	similis
SIMILAR	(homos)	similar, verisimilitude
	homograft, homology,	
	homogeneous	
ALL	παν (gen. παντος)	omnis
	(pan, pantos)	omnibus, omniscient
	panchromatic, pantograph	
ALMOST		paene
		peninsula, penultimate
ALONE	μονος	solum
	(monos)	(abl.) SOLO
	monologue, monoxide	
AUDIBLE	ἀκουστος	
	(acoustos)	
	acoustics	
AUSTERE	αὐστηρος	austerus
	(austeros)	
BAD	1. κακος	malus
	(kakos)	malnutrition, malice
	cacophony, cacodyl	
	2. δυσ-	
	(dys-)	
	dyspeptic, dyspnoea	
BEAUTIFUL	καλος	pulcher
	(kalos)	pulchritude
	calligraphy, kaleidoscope	
BEST	ἀριστος	optimus
	(aristos)	optimism, optimum
	aristocracy, aristogenesis	

110

BIG	μεγας, μεγαλ- (megas, megal-) *megalith, megacycle,* *megalomania*	1. magnus magnanimous, magnitude 2. grandis grandiloquence 3. vastus vast
BITTER	πικρος (pikros) *picrotoxin, picric acid*	acer acrid, acrimonious
BLIND	τυφλος (typhlos) *typhlosole, Typhlops*	CAECUM Caecilia
BLUNT	ἀμβλυς (amblys) *amblyopia, Amblypoda*	1. obtusus obtuse 2. hebes hebetude, hebephrenia
BOTH	ἀμφω, ἀμφι- (ampho, amphi-) *Amphibia, amphicoelous*	ambo ambidextrous, ambivalent
BRIGHT	ἀγλαος (aglaos) *Aglaonema, Aglaophenia*	clarus clarity
BROAD	εὐρυς (eurys) *Eurypterida, eurysome,* *euryhaline*	latus latitude, latifoliate
COLD	1. κρυος (cryos) *cryohydric, cryoscopic* 2. ψυχρος (psychros) *psychrophilic, psychrometer*	1. frigidus frigid, frigorific 2. gelidus gelid 3. algidus algidity, algid, algor
COMMON	κοινος (coenos) *coenocyte, Coenurus*	communis communal, community
COMPACT	πυκνος (pycnos) *pycnic, pycnidiophore*	compactus
CROSSWISE, SLANTING	1. πλαγιος (plagios) *Plagiostomi, plagiotropic* 2. λοξος (loxos) *loxodont, loxodromic*	obliquus oblique

CURVED	1. καμπυλος	1. curvus
	(campylos)	curviserial
	campylotropous,	2. sinuosus
	campylospermous	sinuous, insinuate
	2. κυρτος	
	(cyrtos)	
	cyrtograph, cyrtosis	
	3. ἀγκυλος	
	(ancylos)	
	Ancylostoma	
CUT	τομος	sectile
	(tomos)	
	anatomy, lobotomy	
DEAD	νεκρος	mortuus
	(necros)	mortuary, mortician
	necromancy, necrophilia	
DEEP	βαθυς	profundus
	(bathys)	profound
	bathymetric, Bathycrinus	
DIFFERENT	ἑτερος	diversus
	(heteros)	diverse
	heterogeneous, heterodyne	
DISTANT	τηλε (adv.)	distans
	(tele)	distant
	telephone, telepathy	
DOUBLE	διπλοος	DUPLEX, duplicis
	(diploos)	duplicate, duplicity
	diplococcus, diploblastic	
DRY	ξηρος	siccus
	(xeros)	desiccate, siccative
	xerophilous, xerophyte	
EASY		facilis
		facility, facilitate
ELEGANT	κομψος	elegans
	(compsos)	elegant
	Compsognathus	
EMPTY	κενος	vacuus
	(cenos)	vacuum, evacuate
	cenotaph, kenocis	
ENOUGH		satis
		satisfy, insatiable
EQUAL	ἰσος	aequus
	(isos)	equal, equidistant
	isobar, isosceles	

FEW	ὀλίγος	paucus
	(oligos)	paucity, paucispiral
	Oligochaeta, oligarchy,	
	oligocarpous	
FLAT	πλατυς	planus
	(platys)	plane
	platypus, Platyhelminthes	
FLOWERING	ἀνθηρος	florens
	(antheros)	
	antheridium, anther	
FOOLISH	μωρος	1. stultus
	(moros)	stultifying, stultiloquence
	moronic, oxymoron	2. fatuus
		fatuous, infatuate
FOREIGN	ξενος	1. barbarus
	(xenos)	barbarous, barbarian
	xenophobia, Xenopus	2. peregrinus
		peregrinations
FREE	ἐλευθερος	liber
	(eleutheros)	liberty, liberate
	Eleutheria, Eleutheroblastea	
FULL	πληρης	plenus
	(pleres)	replenish, plenty
	plerocercoid, plerome	
GENERAL	καθολικος	generalis
	(catholicos)	general
	catholic	
GIGANTIC	γιγαντειος	1. ingens
	(giganteios)	2. giganteus
	gigantic	
GOOD	ἀγαθος	BONUS
	(agathos)	bona fide, bonanza
	Agatha, agathism	
HAIRY,	δασυς	hirsutus
ROUGH	(dasys)	hirsute
	Dasyurus, dasyproctid	
HARD	σκληρος	durus
	(scleros)	durable, dura mater
	sclerosis, sclerenchyma	
HARMFUL		nocuus
		nocuous, innocuous
HEALTHY	ὑγιης	sanus
	(hygies)	sanity, sanitary
	hygiene, hygiology	

HEAVY	βαρυς	gravis
	(barys)	gravity, gravid
	barometer, isobar	
HIGH	ἀκρος	altus
	(acros)	altitude, exalt
	acropetal, acrodont	
HINDMOST	ὀπισθε	postremus
	(opisthe)	
	opisthosoma,	
	Opisthobranchiata	
HOLLOW	κοιλος	cavo (fem. cava)
	(coelos)	cavity, concave, VENA CAVA
	Coelenterata, acoelous	
HOLY	ἀγιος	sanctus
	(hagios)	sanctuary, sanctity
	hagiolatry, hagiology	
HOT	θερμος	fervens
	(thermos)	fervent
	thermometer, isotherm	
ILL		aegrotus
		aegrotat
IMMORTAL	ἀμβροτος	immortalis
	(ambrotos)	immortal
	ambrosia, ambrotype	
IMPERFECT	ἀτελης	inchoatus
	(ateles)	inchoate
	atelognathia, atelomyelia	
INDIVIDUAL	ἰδιος	singuli
	(idios)	singularity
	idiosyncrasy, idiom	
INNUMERABLE	μυριος	innumerabilis
	(myrios)	innumerable
	Myriapoda, myriads	
LEAN	ἰσχνος	macer
	(ischnos)	macerate, emaciated
	Ischnochiton	
LEFT		1. laevus
		laevose, laevogyrous
		2. SINISTER
		sinistral, sinistrorse
LITTLE	μικρος	1. parvus
	(micros)	multum in parvo, parviscient
	microscope, micrometer	2. pusillus
		pusillanimous

LONG	1. δολιχος	1. longus
	(dolichos)	longitude, longicaudate
	dolichocephalic,	2. prolixus
	dolichostylous	prolix
	2. μακρος	
	(macros)	
	macroscopic, macronucleus	
MANIFEST	1. δηλος	manifestus
	(delos)	manifest
	Urodela, psychedelic	
	2. φανερος	
	(phaneros)	
	Phanerogam	
MANY, MUCH	πολυς	multi
	(polys)	multifid, multiply
	polygon, polymath,	
	Polyzoa	
MEDICAL	ιατρικος	medicus
	(iatricos)	medical
	paediatrics, iatric	
MIDDLE	μεσος	medius
	(mesos)	medial, mediocre,
	Mesozoic, mesoderm	
MODELLED	πλαστος	
	(plastos)	
	plastic, chloroplast	
MOIST, WET	ὑγρος	humidus
	(hygros)	humid, humidity
	hygroscopic,	
	hygrometer	
MORE	πλειων	PLUS, plure
	(pleion)	plurilocular
	pleiotaxy, pleomorphism	
MOST	πλειστος	maximus
	(pleistos)	maximum, maximal
	Pleistocene	
MOVEABLE		mobili
		automobile
NAKED	γυμνος	nudus
	(gymnos)	denude, nudity
	Gymnosperm, gymnastics	
NARROW	στενος	angustus
	(stenos)	anguish, angustifoliate
	Stenopus, Stenocarpus	

NEAR	πλησιος	1. propinquus
	(plesios)	propinquity
	Plesiosauria, Plesianthus	2. vicinus
		vicinity
NEW	1. καινος	novus
	(caenos)	novel, novice
	caenozoic, Oligocene	
	2. νεος	
	(neos)	
	neolithic, neologism	
ODD,	περισσος	impar
UNEVEN	(perissos)	imparipinnate, imparidigitate
	perissodactyl	
OF EACH	ἀλληλων	
OTHER	(allelon)	
	allelomorph	
OLD	1. παλαιος	1. antiquus
(*ancient*)	(palaeos)	antiquity
	palaeolithic,	
	palaeozoic	
	2. ἀρχαιος	
	archaeos	
	archaic,	
	Archaeopteryx	
(*elderly*)	3. γεραιος	2. senex (*abl.* sene)
	(geraeos)	*senescence*
	geriatric	3. senilis, *senile*
OPPOSITE	ἐναντιος	adversus
	(enantios)	adverse
	enantiomorph, enantioblastic	
OTHER	ἀλλος	1. alius
	(allos)	aliunde, aliquot
	allotropic, allergy	2. alter
		alter ego, alternate
PERFECT,	ἀρτιος	perfectus
EVEN	(artios)	perfect
	Artiodactyla, artiad	
POOR	πενης	PAUPER
	(penes)	
PRIVATE	ἰδιος	1. privatus
	(idios)	private
	idiosyncrasy, idiopathic	2. peculiaris
		peculiarity
PUTRID	σαπρος	putris
	(sapros)	putrid
	saprophyte, Saprolegnia	

QUICK	ταχυς	celer
	(tachys)	accelerate, celerity
	tachygenesis, tachycardia	
REMOTE,	ἐσχατος	remotus
FURTHEST	(eschatos)	remote
	eschatology	
RIGHT		dexter
		dextrose, ambidextrous
ROUGH	τραχυς	
	(trachys)	
	Trachymedusae, trachysoma	
ROUND	στρογγυλος	rotundus
	(strongylos)	rotund, Rotunda
	Strongylus, Strongylo-	
	centrotus	
SCANTY	μανος	exiguus
	(manos)	exiguous
	manometer, manoscope	
SHARP	ὀξυς	acutus
	(oxys)	acute
	oxygen, Amphioxus	
SHORT	βραχυς	brevis
	(brachys)	abbreviate, brief
	brachydactyly, brachycephalic	
SHUT	κλειστος	
	(cleistos)	
	cleistogamous	
SIMILAR	ὁμος	similis
	(homos)	similar
	homology, Homoptera	
SIMPLE	ἁπλοος	simplex
	(haploos)	simplicity
	haploid, Haplosporidia	
SLOW	βραδυς	tardus
	(bradys)	retard, tardigrade
	bradycardia	
SMOOTH	1. λειος	lubricus
	(leios)	lubricate, lubricious
	leiodermatous, leiotrichous	
	2. λισσος	
	(lissos)	
	lissoflagellate	
SOFT	μαλακος	mollis
	(malacos)	Mollusca, emollient
	Malacostraca, Malacocotylea	

SOLID	στερεος	solidus
	(stereos)	solid, consolidate
	stereoscopic, stereogram	
SOLITARY	ἐρημος	solitarius
	(eremos)	solitary
	eremetical, eremurus	
STRAIGHT	1. εὐθυς	rectus
	(euthys)	rectum, rectitude
	Euthyneura, Euthynotus	
	2. ὀρθος	
	(orthos)	
	orthodoxy, Orthoptera	
SWEET	1. γλυκυς	1. dulcis
	(glycys)	dulcet, dulcify
	glycogen, glucose	
	2. ἡδυς	2. suavis
	(hedys)	suave, persuasion
TERRIBLE	δεινος	terribilis
	(dinos)	terrible
	Dinosaur, Dinornis	
THICK	παχυς	1. (con)densus
	(pachys)	density, condense
	pachydermatous, pachymeter	2. grossus
		gross
THIN	1. ἀραιος	tenuis
	(araeos)	tenuous, attenuate, tenuiroster
	araeometer, araeotic	
	2. λεπτος	
	(leptos)	
	Leptostraca, Leptothrix	
TRUE	ἐτυμος	1. verax verae-
	(etymos)	veracious, veracity
	etymology	2. verus
		verism, veritable
TWISTED	στρεπτος	
	(streptos)	
	Streptococcus, Streptomyces	
VAIN	ματαιος	
	(mataeos)	
	mataeotechny,	
	mataeology	
VARIED	αιολος	variatus
	(aeolos)	
	aeolotropic	

VARIOUS	ποικιλος	varius
	(poecilos)	various, variety
	poecilothermic,	
	poecilocyte	
VISIBLE	φανερος	visibilis
	(phaneros)	visible
	Phanerogam, Phanerocephala	
WANDERING	πλανος	
	(planos)	
	planoblast, aplanospore	
WELL (adv.)	εὐ	bene
	(eu)	benediction, beneficial
	euphony, eulogy	
WHOLE	ὁλος	1. integer
	(holos)	integrifolious, integripallial
	holoblastic, holozoic	2. totus
		total
WILD	ἀγριος	ferus
	(agrios)	feral
	agriology	
WISE	σοφος	sapiens
	(sophos)	Homo sapiens, sapient
	Sophist, sophisticated	
WOOLLY	οὐλος	1. lanatus
	(oulos)	lanate
	ulotrichous	2. laniger
		lanigerous

19. IRREGULAR LATIN COMPARISON[1]

ABOVE	supra	superiore	supremo
	suprarenal	*superior*	*supreme*
BAD	malo	peiore	pessimo
	malfunction	*pejorative*	*pessimist*
BEHIND	post	posteriore	postremo
	postgraduate	*posterior*	
BELOW	infra	inferiore	infimo
	infra-red	*inferior*	
BEYOND	ultra	ulteriore	ultimo
	ultra-violet	*ulterior*	*ultimate*
BIG	magno	maiore	MAXIMUM (neut. nom.),
	magnanimous		maximo
		major	*maximal*

[1] Ablative singular masculine case forms of adjectives except where otherwise stated.

GOOD	bono	meliore	OPTIMUM (neut. nom.), optimo
	bonus	*ameliorate*	*optimistic*
IN FRONT	prae	priore	primo
	premolar	*priority*	*primary*
INSIDE	intra	interiore	intimo
	intramolecular	*interior*	*intimate*
MANY	multo	plure	plurimo
	multitude	*plural*	
NEAR	prope	propiore	proximo
	propinquity		*approximate*
OLD	sene	seniore	
	senile	*senior*	
OUTSIDE	extra	exteriore	extremo
	extramural	*exterior*	*extreme*
SMALL	parvo	minore	MINIMUM (neut. nom.), minimo
	parvitude	*minor*	*minimal*
YOUNG	iuvene	iuniore	
	juvenile	*iunior*	

20. VERBS

ANNOUNCE	ἀγγελλω	pronuntio
	(angello)	pronounce
	evangelist, angel	
BE ABLE	δυναμαι	possum, potest
	(dynamai)	possible, impotent
	dynamic, dynast	
BE BORN	γεινομαι	nascor (natus sum)
	(geinomai)	nativity, antenatal,
	progeny	nascent
BELIEVE		credo
		credible, incredulous
BEND	κλινω	flecto (flexi)
	(clino)	flexible, flex
	klinostat, syncline	
BITE	δακνω	mordeo
	dacno	mordant, remorse,
		morsel
BREAK	κλαω	frango (fractum)
	(clao)	fracture, refract
	clasmatocyte, clastic	

BURN	φλεγω (phlego) *phlegmasia, phlegmon*	1. cremo cremation, crematorium 2. incendo (incensum) incendiary, incense
BURST, BREAK	ῥηγνυμι (rhegnymi) *haemorrhage, menorrhagia*	rumpo (ruptum) rupture
CARRY	φορεω (phoreo) *melanophore, xanthophore*	1. fero (latum) aquifer, transfer, translate 2. porto transport, export 3. veho (vexi, vectum) vehicle, vexillum, vector
CHEW	τρωγω (trogo)	rumino Ruminant, ruminate
COOK	πεπτω (pepto) *eupeptic, pepsin*	coquo (coctum) concoct, coctile
COVER	καλυπτω (calypto) *Calyptoblastea, eucalyptus*	tego (tectum) integument, tectorial
CREATE	ποιεω (poieo) *poetry, onomatopoeia, pharmacopoeia*	1. creo creation 2. genero generate
CREEP		repo reptile, repens
CUT	τεμνω (temno) *Temnocephali, temnospondylous*	1. scindo (scissum) scissors, rescind 2. seco (sectum) sectile, secant
DECEIVE	ψευδω (pseudo) *pseudonym, pseudomorph*	fallo (falsum) fallacy, fallible, falsify
DEVOUR	φαγειν (infinitive) (phagein) *phagocyte, entomophagous, phage*	devoro devour
DIP	βαπτω (bapto) *baptism, baptize*	mergo (mersum) merge, emerge, immerse
DISCOVER	εὑρισκω (heurisco) *heuristic, eureka*	invenio (inventum) invent, inventor

DISTRIBUTE, DIVIDE	1. δαιω (daio) geodesy	1. distribuo distribution
	2. κρινω (crino)	2. divido (divisum) divide, division
DROP	σταλαω (stalao) stalactite, stalagmite	stillo still, distillation
DWELL	οἰκεω (oeceo) dioecious, monaecious	1. habito habitat, inhabitant 2. colo arenicolous, saxicolous
FAN	ῥιπιζω (rhipizo) Rhipidium, Rhipidoglossa	ventilo ventilation
FEAR	φοβεομαι (phobeomai) Anglophobe, hydrophobia	timeo timid, timorous
FLOW	ῥεω (rheo) rheostat, diarrhoea	fluo (fluxum) fluid, flux, fluctuate
FRIGHTEN	φοβεω (phobeo) phobia	terreo terrible, deterrent
GOVERN	κρατεω (crateo) plutocratic, democrat	1. impero imperial 2. regno regnant
GRASP	ἁπτω (hapto) haptometer, haptic	1. prehendo comprehend 2. prehenso prehensile
GROW	φυω (phyo) symphysis, hypophysis	1. cresco crescent, increase 2. augeor (auctum) augment, auction
HEAR	ἀκουω (acouo) acoustics, acoumeter	audio audible, audience
HIDE	κρυπτω (crypto) Cryptogam, cryptozoic	occulto occult
KILL	φονευω (phoneuo)	caedo (cecidi) suicide, vermicide, homicide

LIVE	βιοω	vivo
	(bioo)	vivacious, vivisection
	biology, amphibious	
LOOK AT	σκοπεω	specto
	(scopeo)	inspect, spectacle
	telescope, periscope	
LOVE	φιλεω	amo
	(phileo)	amatory, amateur
	philology, Anglophile, entomophilous	
MARRY	γαμεω	nubo (nupsi, nupta sum)
	(gameo)	nubile, nuptial
	polygamy, gamete	
MEASURE	μετρεω	metior (mensus)
	(metreo)	mensuration
	metric, hexameter	
MIX	κεραννυμι	misceo (mixtum)
	(cerannymi)	miscegenation, mixture
	idiosyncrasy	
MOVE	κινεω	moveo (motum)
	(cineo)	move, motion, motor
	cinema, kinetic	
PUTREFY = cause to rot	σηπω, *fut* οηψο (sepo, sepso) *sepsis, antiseptic*	1. putresco putrescent 2. putrefacio putrefy
ROUSE	ὁρμαω	suscito
	(hormao)	resuscitate
	hormone, hormetic	
RUB	τριβω	frico (frictum)
	(tribo)	fricative, friction
	diatribe, tribometer	
RULE	ἀρχω	rego
	(archo)	regal, regulate
	tetrarch, monarch	
SAIL	πλεω	navigo
	(pleo)	navigator, circumnavigate
	pleopod	
SAW	πριω, *fut* πρισω	serro
	(prio, priso)	serrated, serricorn
	prism, prismatic	
SAY, SPEAK	λεγω	dico (dictum)
	(lego)	dictaphone, predict, contradiction

SEEK	ζητεω (zeteo)	1. quaero (quaesitum) question, quest 2. peto centripetal, petition
SELL	πωλεω (poleo) *monopoly*	vendo vendor
SEPARATE	κρινω (crino) *apocrine, endocrine*	separo separate
SERVE	διακονεω (diaconeo) *deacon, diaconal*	servio servant
SHINE	1. λαμπω (lampo) *lamp* 2. φαινω (phaeno) *phenyl*	1. fulgeo fulgent, effulgence 2. splendeo resplendent 3. luceo lucent, translucent
SHOOT (ARROWS)	τοξευω (toxeuo) *toxic, toxaemia*	emitto emit
SHOW	φαινω (phaeno) *phenotype, phenomenon*	monstro demonstrate, monstrance
SLEEP	κοιμαομαι (coemaomai) *cemetery*	dormio dormitory, dormant
SPIN, WHIRL	στροβεω (strobeo) *stroboscope, strobile*	neo
SPLIT	σχιζω (schizo) *schizocarpous, schizophrenia*	findo (fidi, fissum) fission, fissile pinnatifid
SUFFER	ωδινω (odino)	patior (passus) patient, passive, passion
SWIM	νηχω (necho) *Notonecta, nectocalyx*	nato natation, natatorial
TEACH	διδασκω (didasko) *didactic*	doceo docent, docile, doctor

124

THIRST	δυψαω (dipsao) *dipsomania, dipsetic*	1. sitio 2. areo arid
THROW	βαλλω (ballo) *ballistics*	iacio (ieci, iactum) project, trajectory

TUNNEL
  HOLLOW
  OUT
	γλυφω
	(glypho)
	Tyroglyphe, siphonoglyph	

WALK	πατεω (pateo) *peripatetic*	1. ambulo ambulance, perambulate 2. gradior (gressus) gradient, progress
WHIP	μαστιγοω (mastigoo) *Mastigophora,* *Polymastiginae*	flagello flagellate, Flagellata
WORK	ἐργαζομαι (ergazomai) *erg*	1. operor operate 2. laboro labourer
WRITE	γραφω (grapho) *phonograph, photograph*	scribo (scriptum) scribe, script, scripture

PART THREE

Medical and Biological
Greek and Latin Terms

The glossaries exhibited in this section refer to names of organisms or of their parts, and are therefore of relevance chiefly to the needs of those studying biology, pure and applied, including medicine and agriculture. Since the Romans derived their knowledge of anatomy and natural history from the Greek-speaking world, it would lead to avoidable duplication if we followed the plan of the preceding sections of Part Two by listing Greek and Latin equivalents of vernacular terms in parallel columns. Typographical and grammatical conventions in this chapter are otherwise the same. The list of *generic* names for plants and animals is not exhaustive. Most readers who can benefit from this chapter will already be familiar with many of them.

As regards Romanization of Greek words, conventions used in this chapter are the same as those (pp. 61–62) used in Part II.

21. GREEK ZOOLOGICAL AND MEDICAL TERMS

ANUS	πρωκτος	(proctos)	proctodeum, aproctous, Ectoprocta
AORTA	ἀορτη	(aorte)	aortic
APPEARANCE, EYESIGHT	ὀψις	(opsis)	autopsy, Bryopsis, Sauropsida
ARM	βραχιων	(brachion)	brachial
ARTERY	ἀρτηρια	(arteria)	arterial
BACK	νωτον	(noton)	notochord, notopodium, Notostraca
BACKBONE	ῥαχις	(rhachis)	rachitis, rachitomous, Rachitomi
BEARD	πωγων	(pogon)	Ophiopogon, Pogonophora
BELLY	γαστηρ	(gaster)	gastric, epigastric, Gasteromycetes
BILE	χολη	(chole)	glycocholate, melancholia
BLADDER, BAG	κυστις	(cystis)	cystitis, nematocyst
BLOOD	αἱμα	(haema)	haemal, haemoglobin, haemocyanin
BODY	σωμα	(soma)	somatic, centrosome, Pyrosoma
BONE	ὀστεον	(osteon)	osteology, periosteal
BOWELS	σπλαγχνα	(splanchna)	splanchnic, splanchnopleure
BRAIN	ἐγκεφαλος	(encephalos)	mesencephalon, encephalitis

129

BREAST	στερνον	(sternon)	*sternal, sternocostal*
BREATH	πνευμα	(pneuma)	*pneumatic, pneumatometer*
BUTTOCKS	πυγη	(pyge)	*pygostyle, pygal, pygidium*
CALF (of leg)	γαστροκνημη	(gastrocneme)	*gastrocnemius*
CARTILAGE	χονδρος	(chondros)	*Chondrostei, Chondrichthyes*
CHEEK	1. παρεια	(pareia)	*parietal*
	2. γενυς	(genys)	*genyplasty*
CHEST	στηθος	(stethos)	*stethoscope, stethograph*
CLOT	θρομβος	(thrombos)	*thrombosis, thrombocyte*
COMB, CREST	λοφος	(lophos)	*lophodont, Lophopus, Lophogaster*
CORNER (of eye)	κανθος	(canthos)	*epicanthial, canthoplasty*
CUTTING OUT	ἐκτομη	(ectome)	*thyreodectomy, hypophysectomy*
DIGESTION	πεψις	(pepsis)	*pepsin, eupeptic*
DISCHARGE	πυος	(pyos)	*pus, pyogenic*
DRUG	φαρμακον	(pharmacon)	*pharmacist, pharmacology*
DUNG	κοπρος	(copros)	*coprolite, coprophagous*
EAR	οὖς, ὠτος	(ous, otos)	*periotic, otolith, otocyst*
EGG	ὠον	(oon)	*oogenesis, oogonium, oospore*
EMBRYO	ἐμβρυον	(embryon)	*embryonic, polyembryony*
EYE	1. ὠψ	(ops)	*myopia, pyrope*
	2. ὀμμα, -ατος	(omma, -atos)	*ommatidium, Ommastrephes*
	3. ὀφθαλμος	(ophthalmos)	*ophthalmic, ophthalmoscope*
EYEBALL	γληνη	(GLENE)	*glenoid*
EYEBROW	ὀφρυς	(ophrys)	*Actinophrys, Ophryocystis, Ophrytrocha*
EYELID	βλεφαρον	(blepharon)	*Monoblepharis, Blepharipoda*
FEATHER	πτιλον	(ptilon)	*coleoptile, Trichoptilum*
FEVER	πυρετος	(pyretos)	*antipyretic, pyrexia*
FIN	πτερυγιον	(pterygion)	*archipterygium, actinopterygial*
FINGER, TOE	δακτυλος	(dactylos)	*hexadactyl, polydactyly, pterodactyl*
FLESH	1. κρεας	(creas)	*creatine, creatinine, pancreas*
	2. σαρξ, σαρκος	(sarx, sarcos)	*perisarc, sarcoma*
FOOT	πους, ποδος	(pus, podos)	*Amphipoda, Platypus, Lycopodium*
FOOTPRINT	ἰχνος	(ichnos)	*Ichnotropis, ichnology*
GILLS	βραγχια (plur.)	(branchia)	*branchial, Branchiopoda, Branchiura*
GLAND	ἀδην	(aden)	*adenoid, adenuma*
GULLET	λαρυγξ	(LARYNX)	*laryngeal, laryngitis*

130

GUT	ἐντερον	(enteron)	enteritis, coelenterate, mesentery
HAND	χειρ	(chir)	Chiroptera, chiropodist
HAIR	θριξ, τριχος	(thrix, trichos)	Polytrichum, Trichina, Ophiothrix
HEAD	κεφαλη	(cephale)	acephalic, Cephalopoda
HEALTH	ὑγιεια	(hygieia)	hygiene, hygienic
HEART	καρδια	(cardia)	cardiac, cardiogram
HEEL	πτερνα	(pterna)	Litopterna
HORN	κερας, -ατος	(ceras, -atos)	keratin, rhinoceros
JAW	γναθος	(gnathos)	gnathite, prognathous, Gnathobdella
JOINT	ἀρθρον	(arthron)	Arthropoda, Xenarthra, arthritis
JOINT (of toe or finger)	φαλαγξ	(phalanx)	phalanges, phalangeal
KIDNEY	νεφρος	(nephros)	nephritis, nephridium, mesonephros
KNUCKLE	κονδυλος	(condylos)	condyle, Condylarthra
LEG	σκελος	(scelos)	isosceles, scelalgia
LIP	χειλος	(cheilos)	Chilognatha, Chilopoda
LIVER	ἡπαρ, ἡπατος	(hepar, hepatos)	hepatic, hepatitis
LUNGS	πνευμων	(pneumon)	pneumonia, pneumococcus
MADNESS	μανια	(MANIA)	maniac, hypomania
MANE, long hair	χαιτη	(chaite)	Polychaeta, Chaetognatha, Chaetocladium
MEMBRANE	ὑμην	(HYMEN)	Hymenoptera, Hymenomycetes
MOUTH	στομα, ατος	(stoma, -atos)	stomata, Gnathostomata, Bdellostoma
MUSCLE, mouse	μυς, μυος	(mys, myos)	myomere, myotome, myocardium
NAIL, CLAW	ὀνυξ, ὀνυχος	(ONYX, onychos)	Onychophora, Onychomonas
NERVE, TENDON	νευρον	(neuron)	neural, neurosis
NOSE	ῥις, ῥινος	(rhis, rhinos)	rhinitis, rhinoceros, antirrhinum
NUMBNESS	ναρκη	(narce)	narcosis, narcotic
OESOPHAGUS	οἰσοφαγος	(oesophagos)	oesophagus, oesophagitis
OPENING (of stomach)	στομαχος	(stomachos)	stomach
PAIN	ἀλγος	(algos)	analgesic, neuralgia, algedonic
PENIS	φαλλος	(phallos)	phallic
PHLEGM, MUCUS	μυξα	(myxa)	Myxomycetes, Myxococcus, Myxosporidia

PULSE	σφυγμος	(sphygmos)	sphygmoid, sphygmomanometer
REGIMEN	διαιτα	(diaeta)	diet, dietetics
REMEDY	ἀκος	(acos)	autacoid, acology
SCALE	λεπις, λεπιδο-	(lepis, lepido-)	Lepidoptera, Lepidostei, Osteolepis
SEA-SICKNESS	ναυσια	(nausia)	nauseating, nausea
SHELL	1. ὀστρακον	(ostracon)	Ostracoda, Conchostraca, Entomostraca
	2. κογχη	(conche)	conchology, conchite
SIDE, RIB	πλευρα	(pleura)	pleural, pleurocentrum, pleurisy
SKIN	1. δερμα	(derma)	epidermis, mesoderm, dermatitis
	2. χρως, χρωτος	(chros, chrotos)	Chrotella
SKIN, LEATHER	χοριον	(chorion)	chorion, chorionic, choroid
SKULL	κρανιον	(cranion)	cranial, Craniata, chondrocranium
SMELL, sense of	ὀσφρα	(osphra)	osphradium
SNOUT	ῥυγχος	(rhynchos)	rhynchota, Rhynchocephalia, Rhynchobdellida
SOLE	πελμα	(pelma)	Pelmatozoa
SPASM	σπασμος	(spasmos)	spasmodic
SPLEEN	σπλην	(splen)	splenetic, splenalgia
STING	κεντρον	(centron)	Centronotus
SWELLING	οἰδημα	(oedema)	oedema
SYMPTOM	συμπτωμα	(symptoma)	symptomatic, symptomatology
TAIL	1. οὐρα	(ura)	urostyle, Ophiura, Anura
	2. κερκος	(cercos)	cercaria, Xiphocercus
TALON	χηλη	(chele)	chela, chelate, chelicera
TEAT	θηλη	(thele)	thelin, epithelium
TENDON	τενων	(tenon)	tenology, tenotomy
TESTICLE	ὀρχις	(orchis)	cryptorchid, orchitis, orchotomy
THIGH, HIP	ἰσχιον	(ischion)	ischial, ischiopodite
THROAT	1. βρογχος	(bronchos)	bronchi, bronchitis
	2. φαρυγξ	(pharynx)	glossopharyngeal, Pharyngobranchii
	3. λαιμος	(laemos)	phylactolaematous
TONGUE	γλωσσα	(glossa)	hypoglossal, epiglottis, Ophioglossum
TOOTH	ὀδους, ὀδοντος	(odous, odontos)	Odontophore, thecodont, Odontoceti
TUBERCLE	χαλαζα	(chalaza)	chalaza, chalazogamic
URINE	οὐρον	(uron)	uric, urea, hippuric

VEIN	φλεψ, φλεβος	(phlebs, phlebos)	phlebitis, phlebotomy
VERTEBRA	σπονδυλος	(spondylos)	diplospondylous, Spondylus
VOMIT	εμετος	(emetos)	emetic, emetology
WINDPIPE	τραχεια	(tracheia)	tracheal, tracheate, tracheide
WING	πτερον	(pteron)	Aptera, Hymenoptera, Neuroptera
WOOL	1. πιλος	(pilos)	Pilochrota, Pilobolus
	2. εριον	(erion)	Eriophyes, Eriocaulon, Eriospermeae
WOUND	τραυμα	(TRAUMA)	trauma, traumatic
WRIST	καρπος	(carpos)	carpal, metacarpal
YOLK	λεκιθος	(lecithos)	lecithin, alecithal

22. GREEK NAMES OF ANIMALS

ANIMAL	ζωον	(zoon)	zoology, spermatozoon
ANT	μυρμηξ, μυρμηκος	(myrmex, myrmecos)	myrmecology, myrmecophagous, Myrmecinae
APE	πιθηκος	(pithecos)	pithecanthropus, Cercopithecus
BAT	νυκτερις	(nykteris)	
BEAR	αρκτος	(arctos)	Arctic, Arcturus
BEAVER	καστωρ	(castor)	
BEAST	θηρ	(ther)	Theromorpha, Megatherium
BEE	μελισσα	(melissa)	
BEETLE	κανθαρος	(cantharos)	cantharis
BIRD	ὀρνις, ὀρνιθος	(ornis, ornithos)	ornithologist, Notornis
BUG	κορις	(coris)	Coreidae
BULL	ταυρος	(tauros)	Minotaur, taurine
BUTTERFLY	ψυχη	(psyche)	Psychidae
CAMEL	καμηλος	(camelos)	Camelidae
CATERPILLAR	καμπη	(campe)	Campanotus
COCK	ἀλεκτρυων	(alectryon)	alectryomachy, alectryomancy
CRAB	καρκινος	(carcinos)	carcinology, carcinoma
CROCODILE	κροκοδειλος	(crocodeilos)	
CROW	κοραξ, κορακος	(corax, coracos)	coraciiform, coracoid
CUCKOO	κοκκυξ	(coccyx)	coccygeal
CUTTLEFISH	σηπια	(sepia)	sepiolite, sepiostaire
DOG	κυων, κυνος	(cyon, cunos)	Cynognathus, cynophobia
ELEPHANT	ἐλεφας	(elephas)	elephantiasis
FISH	ἰχθυς	(ichthys)	ichthyornis, ichthyosaur
FLEA	ψυλλα	(psylla)	Psyllidae
FROG	βατραχος	(batrachos)	Batrachian
GLOW-WORM	λαμπυρις	(lampuris)	lampyrine

GOAT	τραγος	(tragos)	Tragopan, Tragus, Tragulidae
GOOSE	χην	(chen)	chenopod
HARE	λαγως	(lagos)	Lagomorpha, lagophthalmus
HEDGEHOG	ἐχινος	(echinos)	Echinodermata, Echinococcus
HORSE	ἱππος	(hippos)	Hippopotamus, hippodrome
INSECTS	ἐντομα (plur.)	(entoma)	entomology, entomophagous
LEECH	βδελλα	(bdella)	bdellatomy, Pontobdella, Bdellostoma
LION	λεων	(leon)	leonine, leopard
LIZARD	σαυρα	(saura)	sauropod, Dinosaur, Ichthyosaur
LOBSTER	ἀστακος	(astacos)	Astacus, astacolite
MACKEREL	σκομβρος	(scombros)	scombroid, Scomber
MITE	ἀκαρι	(acari)	acaroid, acariasis, Acaridae
MONKEY	κερκοπιθηκος	(cercopithecos)	cercopithecoid
MOUSE	μυς, μυος	(mys, myos)	Myosotis, myosin, Myomorpha
OCTOPUS	πολυπους	(polypos)	polyp, polypidom
OSTRICH	στρουθος	(struthos)	Struthio, struthioid
OWL	γλαυξ	(glaux)	
OX	βους	(bous)	buffalo, bugloss, bulimia
OYSTER	ὀστρεον	(ostreon)	Ostreidae, ostreiculture
PARROT	ψιττακη	(psittace)	psittacosis, psittaceous
PARTRIDGE	περδιξ	(perdix)	Perdix
PHEASANT	φασιανος	(phasianos)	Phasianidae
PIGEON	περιστερα	(peristera)	peristerite, peristeronic
PORCUPINE	ὑστριξ	(hystrix)	Hystrix, Hystricomorpha
PORPOISE	φωκαινα	(phocaena)	Phocaena
REPTILE	ἑρπετον	(herpeton)	herpetology, herpetofauna
SALAMANDER	σαλαμανδρα	(salamandra)	Salamandridae
SCORPION	σκορπιος	(scorpios)	Scorpionida
SEAL	φωκη	(phoce)	Phoca, phocodont
SHARK	σελαχος	(selachos)	selachian
SHELLFISH	κογχος	(conchos)	conch, conchology
SHRIMP	καρις	(caris)	Caridea, Caridina
SILKWORM	βομβυξ	(bombyx)	Bombyx
SNAIL	κοχλιας	(cochlias)	cochlea
SNAKE	ὀφις	(ophis)	Ophidia, Gymnophiona
SPIDER	ἀραχνη	(arachne)	arachnid, arachnoid
SPONGE	σπογγια	(spongia)	Demospongiae, Spongidae
SQUIRREL	σκιουρος	(sciuros)	Sciurus, Sciuromorpha
STORK	πελαργος	(pelargos)	Pelargonium
SWAN	κυκνος	(cycnos)	cygnet, Cygnus
TIGER	τιγρις	(tigris)	Felis tigris
TIMBERWORM	τερηδων	(teredon)	teredo
TOAD	φρυνη	(phryne)	Phrynosoma

TORTOISE	χελωνη	(chelone)	*Chelonia*
WHALE	κητος	(cetos)	*Cetacea, spermaceti*
WOLF	λυκος	(lycos)	*Lycognathus, lycanthropy*
WORM	1. ἑλμις, ἑλμινθος	(helmis, helminthos)	*Platyhelminthes, Nemathelminthes, helminthology, helminthiasi*
	2. σκωληξ	(scolex)	*scoleciform, scolecoid, Scolex*

23A. SOME GREEK NAMES OF PLANTS[1]

ANEMONE	ἀνεμωνη	(anemone)
ARTICHOKE	κιναρα	(cinara)
ASPARAGUS	ἀσπαραγος	(asparagos)
CABBAGE	κραμβη	(crambe)
CEDAR (or JUNIPER)	κεδρος	(cedros)
CYPRESS	κυπαρισσος	(cyparissos)
DAFFODIL	ναρκισσος	(narcissos)
FIG	συκον	(sycon)
GRASS	1. ἀγρωστις	(agrostis)
	2. ποα	(poa)
HEATH	ἑρεικη	(ereice)
HELLEBORE	ἑλλεβορος	(helleboros)
HYACINTH	ὑακινθος	(hyacinthos)
HYSSOP	ὑσσωπος	(hyssopos)
IRIS	ἱρις	(iris)
MINT	μινθα	(mintha)
MULBERRY	μορεα	(morea)
MUSTARD	σιναπι	(sinapi)
ORCHID	ὀρχις	(orchis)
PEA	πισος	(pisos)
PEPPER	πεπερι	(peperi)
PLANE TREE	πλατανος	(platanos)
RADISH	ῥαφανις	(rhaphanis)
SAFFRON	κροκος	(crocos)
THYME	θυμος	(thymos)
VINE	ἀμπελος	(ampelos)
CRESS	καρδαμον	(cardamon)

23B. OTHER GREEK BOTANICAL TERMS

BERRY, GRAIN	κοκκος	(coccos)	*Pleurococcus, Diplococcus*

[1] Many of these have passed into use as names of genera. Some however, do not tally with current vernacular terms cited as equivalent. Thus *Cardamine* (lady's smock) though like cress a Crucifer is not edible. Autumn Crocus (*Iridaceae*) is the source of the dye *saffron*. Meadow Saffron (*Liliaceae*), which superficially resembles it, is not.

135

BOUGH, BRANCH	κλαδος	(clados)	Cladophora, phylloclade
BUD	βλαστος	(blastos)	blastoderm, hypoblast
BUNCH of grapes	1. βοτρυς	(botrys)	Botryllus, Botrydium
	2. σταφυλη	(staphyle)	Staphylococcus, staphylinid
CLUSTER of flowers	κορυμβος	(corymbos)	corymb, Corymbocrinus
CONE	κωνος	(conos)	conifer, conidiospores
FERN	πτερις, πτεριδο-	(pteris, pterido-)	Pteridophyta, Pteris
FLAX, LINEN	λινον	(linon)	linen, lineic
FLOWER	1. ἀνθος	(anthos)	Helianthus, Anthozoa
	2. ἀνθεμον	(anthemon)	Chrysanthemum
FRUIT	καρπος	(carpos)	pericarp, syncarpous
HEMLOCK	κωνειον	(coneion)	coniine, Conium
HERB, PLANT	βοτανη	(botane)	botany, botanical
LEAF	φυλλον	(phyllon)	mesophyll, phyllode
LILY	κρινον	(crinon)	Crinoidea, crinid
MOSS	βρυωνη	(bryone)	Bryophyta, Dinobryon
MUSHROOM	μυκης	(myces)	Oomycetes, mycetozoa
NETTLE	1. κνιδη	(cnide)	cnidocil, cnidoblast
	2. ἀκαληφη	(acalephe)	Acalephae
NIGHTSHADE	στρυχνος	(strychnos)	strychnine
NUT,	καρυον	(caryon)	Caryophyllaceae, Caryopsis, karyomere
PEAR	ὀγχνη	(onchne)	Onchnesoma
PETAL	πεταλον	(petalon)	polypetalous, sympetalous
PLANT	φυτον	(phyton)	holophytic, phytology, Spermatophyta
REED	καλαμος	(calamos)	Calamoichthyes, Calamites, calamary
ROOT	ῥιζα	(rhiza)	rhizome, mycorhiza, Rhizopoda
ROSE	ῥοδον	(rhodon)	rhododendron, rhodopsin, rhodium
SEAWEED	φυκος	(phycos)	Phycomycetes, Rhodophyceae
SEED	1. σπερμα	(sperma)	spermatozoa, Gymnosperm
	2. σπορος	(sporos)	sporocyst, Sporozoa
SHOOT	1. κλων	(clon)	clone
	2. θαλλος	(thallos)	thallogenous, Thallophyta
SPINE, THORN	ἀκανθα	(acantha)	Pyracantha, Acanthocephali
STALK	καυλος	(caulos)	cauline, cauliflower
STICK	ῥαβδος	(rhabdos)	rhabdite, Rhabdocoelida
TENDRIL, SPIRAL	ἑλιξ, ἑλικος	(helix, helicos)	helicoid, helicopter

136

TREE	1. δενδρον	(dendron)	*rhododendron, dendrite*
	2. δρυς	(drys)	*dryad, Dryopithecus*
TRUNK	κορμος	(cormos)	*corm, cormogen, cormophyly*
WOOD,	1. ὕλη	(hyle)	*hylophagous, Hylobates*
TIMBER	2. ξυλον	(xylon)	*xylem, xylonite, xylophone*
YEAST,	ζυμη	(zyme)	*enzyme, zymotic, zymase*
LEAVEN			

24. LATIN ANIMAL BODY AND MEDICAL TERMS

Elsewhere our lists have cited the ablative singular form of the Latin noun. Such is the singular form preserved in Spanish, Portuguese and Italian. It also exposes the stem, when the dictionary form (e.g. nominative *corpus*, ablative *corpore* as in *corporal*) fails to do so. Many Latin nouns, however, especially as names of animals and plants, have come into international usage in their dictionary form, and, as such, appear in capital letters in lists 24–26 below.

ABDOMEN		*abdominal*
armpit	AXILLA	*axillary*
back	1. dorso	*dorsal, endorse*
	2. tergo	*tergiversation, tergite, tergal*
beak	rostro	*rostral, tenuiroster*
beard	barba	*barber, barbellate*
belly	ventre	*ventral, ventriloquism, ventricular*
big toe	HALLEX	
bladder	vesica	*vesicle, vesiculitis*
blood	1. sanguine	*sanguinary, ensanguine*
	2. CRUOR	*cruorin*
body	CORPUS	
	(pl. CORPORA)	*corporal, corporate*
bone	Os	*osseous, ossification*
brain	CEREBRUM	*cerebral, cerebellum*
breast	MAMMA	*mammary, Mammalia*
breastbone	1. STERNUM	*sternal, sternebra*
	2. pectore	*pectoral, expectorate*
breath	1. anima	*animate, inanimate*
	2. spiritu	*spiritual*
bristle	S(A)ETA	*setaceous, setiform, setirostral*
cheek	1. bucca	*buccal*
	2. GENA	*genal, genial*
	3. MALA	*malar*
claw, hoof	1. UNGUES	*unguiculate*
	2. UNGULA	*ungulate, unguligrade*
cough	TUSSIS	*tussive, pertussis*
curl	CIRRUS	*cirrate, Cirripedia, cirrocumulus*

disease	morbo	morbidity, morbillous, morbific
drug	medicamento	medicament
ear	aure	aural, auricula
egg	OVUM	oviparous, oviduct
excrement	1. stercore	stercoraceous, stercoricolous
	2. EXCREMENTA	
eye	1. oculo	ocular, oculist
	2. OCELLUS	ocellated, Blennius ocellaris
eyebrow	palpebra	epalpebrate
eyelash	CILIUM, pl. CILIA	ciliary, ciliolate, Ciliophora, Ciliata
feather, fin	1. pluma	plume, plumose, Plumularia
	2. penna	pennaceous, Pennatula
	3. PINNA	pinnate, Pinnipedia
fever	febre, -i	febrile, febrifacient
finger, toe	digito	digital, digitigrade
foot	pede	pedal, centipede, millipede
footprint, trace	vestigio	vestige
forehead	fronte	frontal
fur	PILUS	depilatory, pilose
giddiness	VERTIGO	
gum	gingiva	gingivitis, gingival
hair	1. capillo	capillary, capilliform, capillarimeter
	2. crine	criniparous, Criniger
hand	manu	manual, manufacture, manuscript
head	capite	decapitate, capitulum
health	sanitate	sanitation, sanity
hearing	auditione	audition, auditory, audiometer
heart	corde	cordiform, cordate
heel, ankle	1. TALUS	talon, talaria, talipes
	2. CALX	calcigrade, calciform, calcaneum
	3. CALCANEUM	calcaneal
hip	COXA	coxal, coxopodite, coxocerite
horn	cornu	cornuate, Capricorn, cornucopia
jaw	1. MAXILLA	maxillary, maxilliped, maxilliform
	2. MANDIBULA	mandible, mandibulate
joint	ARTICULUS	article
kidney	rene	renal, reniform, adrenalin
knee	genu	genuflect, genuclast, genual
leg	CRUS, pl CRURA	crural
lip	1. LABIUM	labial, labiatiflorous, Labiatae
	2. LABRUM	labral, labret
lung	pulmone	pulmonary, pulmoniferous, pulmobranchial
madness	insania	insane, insaniate
medical	medicus	medical

medicine	medicina	*medicinal*
mouth	Os, ore	*oral, orarium*
muscle	musculo	*muscular, musculation*
neck	1. Cervix	*cervical, cervicitis*
	2. collo	*collar, accolade*
nerve	nervo	*nervous, nervine, nervation*
nose	naso	*nasal, nasturtium, nasiform, nasute*
palm	palma	*palmate, palmiped, palmistry,*
		palmigrade
phlegm	Mucus	*mucilage*
pulse	pulso	*pulse, propulsion, impulsive, pulsatile*
refuse	Faeces (*pl*)	
remedy	remedio	*remedial, irremediable*
rib	costa	*costal, unicostate, intercostal*
scale	1. Lamina	*laminate, laminable*
	2. Squama	*squamous, squamiform, desquamate*
shell	testa	*testaceous*
shoulder	Humerus	*humeral, humerocubital*
shoulderblade	Scapula	*scapular, scapulary, scapulet(te)*
side	latere	*lateral*
sight	visione	*vision, visual*
skin, hide	1. Cutis	*cuticle, subcutaneous*
	2. pelle	*pellagra, pellicle*
	3. Corium	*excoriate, coriaceous*
skull	Calvaria	*Calvary* (= Golgotha, i.e. place of skulls)
sleep	1. somno	*somnolent. insomnia, somnipathy*
	2. sopore	*soporific, soporiferous*
sole	1. Planta	*plantigrade, plantar*
	2. Solea	*sole, soleaform*
spine, backbone	spina	*spinal*
sweat	sudore	*sudorific, sudoriferous, sudoriparous*
tail	cauda	*caudal, Caudata, caudad, caudiform*
tear	lacrima	*lacrimal, lachrymose*
temple	tempore	*temporal, temporo-malar*
thigh	Femur, -ore	*femoral*
throat	1. Rumen	*ruminant, ruminate*
	2. Gula	*gular, gulist*
	3. Fauces	
thumb	Pollex	
tongue	lingua	*lingual, linguist*
tooth	dente	*dental, dentist, dentate*
touch (sense of)	[con]tactu	*contact, tactile, tact, tactor*
urine	urina	*urinary, urinal, uric, urinate*

vein	VENA	*venous, venosity, venation*
VERTEBRA		*vertebrate*
VISCERA		*eviscerate, visceral*
wart	VERRUCA	*verruciferous, verruculose*
wing[1]	ALA	*alary, alate, alation, aliform*
womb	UTERUS	*uterine, uterogestation*
wool	lana	*lanolin, lanigerous, laniferous*
wound	vulnere	*vulnerable, vulnerate*
wrinkle	RUGA	*rugate, rugulose, rugosity, corrugated*

25. SOME LATIN NAMES OF ANIMALS

Animal	1. ANIMAL	*animalism*
	2. fera	*feral*
	3. pecore	*Pecora*
Ant	FORMICA	*formic acid*
Bat	VESPERTILIO	
Bear	URSUS, URSA (she-bear)	*ursine, Ursa Major, Ursa Minor*
Bee	APIS	*apiary, apivorous, apiculture*
BEETLE	scarabeo	*scarab*
Bird	AVIS, AVES (pl.)	*avine, aviary, avifauna, aviation*
Bull	TAURUS	*Minotaur, taurine*
BUTTERFLY	PAPILIO	*Papilionidae*
CARP	CYPRINUS	*Cyprinidae*
Cat	FELIS	*feline*
Cockroach	BLATTA	*Blattidae*
Cow	vacca	*vaccine, vaccination*
Crab	1. CANCER	*cancroid, canker*
	2. PAGURUS	
CUCKOO	1. CUCULUS	*Cuculiformes*
	2. COCCYX	*coccygeal*
Dog	CANIS	*canine, Canicula, Canis Major*
Donkey	1. ASINUS	*Equus asinus, asinine*
	2. ONAGER	
Duck	ANAS, anate	*Anatidae*
EAGLE	AQUILA	*aquiline*
Eel	ANGUILLA	
Fish(es)	PISCES (pl.)	*pisciform, piscivorous*
Flea	PULEX	*Pulicidae*
Fly	MUSCA	*Muscidae*
Fowl (Domestic)	GALLUS	*Gallinae*
Fox	VULPES	*vulpine*
Frog	RANA	*Ranidae, ranunculus*
Gnat	CULEX	*Culicidae*

[1] Lateral wing, like petals of the pea, etc.

Goat	capro	*capric acid, caprine, Capricorn*
Goose	ANSER	*anserine*
Grasshopper	GRYLLUS	*Gryllidae*
Hare	LEPUS, lepore	*Leporidae*
Hen	GALLINA	*gallinaceous*
Horse	1. EQUUS	*equine, equitation, Equisetum*
	2. CABALLUS	*E. caballus, caballero*
Lamb	agnus	*Agnus castus, Agnus dei*
Leech	HIRUDO	*Hirudinea*
Lion	LEO, -onis	*Felis leo, leonine*
Lizard	LACERTA	*Lacertidae, lacertiform*
Locust	LOCUSTA	*Locustidae*
Louse	PEDICULUS	*Pediculina*
Mole	TALPA	*Talpidae*
Monkey	SIMIA	*simian*
Mouse	MUS, mure	*murine*
Owl	1. STRIX, strigis	*Strigiformes*
	2. BUBO, bubonis	
Ox	Bos, bovis	*bovine*
Oyster	OSTREA	
Partridge	PERDIX	
PIG	1. SUS	*swine*
	2. porco	*pork, porcine, porcupine*
Pigeon	COLUMBA	*columbiformes*
Rabbit	CUNICULUS	*Lepus cuniculus*
Salmon	SALMO, -onis	*Salmonidae*
Sawfish	PRISTIS	
Shark	SQUALUS	
Sheep	OVIS	*ovine*
Snail	COCHLEA	
Snake	1. ANGUIS	
	2. serpente	*serpentine*
	3. COLUBER	
	4. VIPERA	
Sole	SOLEA	
Sparrow	PASSER	*passerine*
Spider	ARANEA	*Araneae*
Stork	CICONIA	*Ciconiiformes*
Swallow	HIRUNDO	*Hirundinidae*
Toad	BUFO, -one	*Bufonidae*
Wasp	VESPA	*Vespidae*
Weasel	MUSTELA	*Mustelidae*
Whale	BALAENA	*Balaenidae*
Wolf	LUPUS	*Canis lupus, lupine*
Worm(s)	VERMES (*pl.*)	*vermiform, vermicide*

apple	malo	*malic acid*
ash	FRAXINUS	*froxin, fraxinella*
bark	CORTEX, cortice	*cortical, corticate, corticolous*
barley	HORDEUM	*hordein*
bean	FABA	*fabiform, fabella*
beech	FAGUS	
berry	BACCA	*bacciferous, baccivorous*
branch	ramo	*ramiferous, ramification*
bud	gemma	*gemmate, gemmiparous*
bunch of grapes	uva	*uvea, uvula, Uvularia*
cabbage	BRASSICA	
carrot	DAUCUS	
chestnut	CASTANEA	*castaneous, castanet*
fig	FICUS	*ficoidal*
fir	ABIES	*abietic, abietin, abietene*
	PICEA	
flower	flore	*flora, nudiflorum*
foliage	fronde	*frond, frondescence, frondiferous*
fruit	fructu	*fructiferous, fructescent*
	fruge	*frugivorous, frugiferous*
	pomo	*pome, pomegranate, pomander*
garlic	allio	*allyl, alliaceous*
grass	gramine	*Graminaceae, graminiferous*
hemp	CANNABIS	*cannabin*
herb	herba	*herbal, herbaceous*
leaf	FOLIO	*foliage, foliaceous*
lettuce	LACTUCA	
mulberry	MORUS	*morula, morulation, morulit*
mushroom	FUNGUS	*fungiform, fungicide*
nettle	URTICA	*urticaria, urticate*
oak	QUERCUS	*quercetin, quercitannin, quercite*
oats	AVENA	*avenaceous, avenage, aveniform*
olive	OLEA	*oleaginous, oleaceous*
pea	PISUM	*pisiform, pisolite*
peach	PERSICA	*persicaria, persico*
pear	PYRUS	*pyruline, pyriform*
pine	PINUS	*pinoleum, pinite, pinic*
plum	PRUNUS	*pruniferous, prunello, prune*
poplar	POPULUS	*populin*
reed	CANNA	*cane, cannula, Cannaceae*
rice	ORYZA	*oryzivorous*
root	RADIX, radice	*radical, radicolous, eradicate*
rye	SECALE	

seaweed	Alga	*algal*
seed	Semen, *abl.* semine	*seminiferous, insemination*
stem, stalk	1. Caulis	*cauliflower, caulescent, caulicule*
	2. stipula	*stipulate, stipuliform*
tree	Arbor	*arboreal, arborescent*
trunk	1. Truncus	*truncate*
	2. Stirps	*stirpiculture*
turnip	napo	*napiform*
vegetable	(h)olere	*oleraceous*
vine	1. vinea	*vine, vinous*
	2. Vitis	*viticulture*
wheat	Triticum	*triticin*
willow	Salix	*salicaceous, salicylic*

Epilogue
Preserving our Heritage

We have seen how and why scientific workers in the closing years of the eighteenth century began to create a new vocabulary suitable for international use. Before then, they had been content to adapt to their requirements words of everyday speech or words of the auxiliary medium used by scholars for instruction and written communication, i.e. Latin in Western Christendom. There were three ways of adapting vernacular words to the end in view:

(i) to impose on the individual word a definition more restricted than, but included in, its more usual meaning;

(ii) to give a word a new meaning suggested by, but not included in, its customary connotation;

(iii) to make a compound by combining in a new way words in everyday use.

The physicist's definition of *velocity*, in contradistinction to *speed*, is an example of the first. The Newtonian use of the word *force* illustrates the second, as does likewise an instruction to *earth* a terminal. To speak of a storage battery as *overcharged* exemplifies the third.

The disadvantage of borrowing and adapting vernacular words for technical terms is that it generates new opportunities for self-deceptive double talk. Economics is hag-ridden with a vocabulary of this sort. With more disastrous consequences to natural science, the same is true of statistical theory. In the domain of statistics, such terms as *significance* and *confidence* both invite and endorse mental confidence tricks. An isolated early example of the verbal technique which first gained ground on the threshold of the nineteenth century will bring sharply into focus the merit of a different procedure.

In the mid-seventeenth century, Van Helmont introduced *gas* to replace the word *spirits*, as in *spirits of salt* (HCl). Gas was in fact a Dutch rendering of the Greek word Romanized as *chaos*. Derived as such from a dead language with which few chemists of his time were familiar, *gas* had no semantic associations apt to befog discussion of the newly discovered *third state of matter*.

In the context of this innovation the alternative term was still redolent with Aristotelian residues. In the Aristotelian Weltanschauung, lucidly expounded by St Paul (1 Cor. 15), things were either terrestrial and material or spiritual and celestial. Because things seek the place where they belong, material bodies fall to the earth and spiritual bodies rise to heaven. To account for the ascent of spiritual bodies (i.e. bodies lighter than air) Aristotelians endowed them with the opposite of gravity, i.e. levity. Its ghost lingered on as phlogiston long after Galileo had established the modern principle of terrestrial gravitation.

The substitution of *gas* for *spirits* draws attention to an outstanding characteristic of the reforms initiated by Linnaeus and Lavoisier. In creating new terms drawn from one or other of two *dead* languages, it was possible to endow them with unique meaning wholly devoid of irrelevant overtones. In the vernacular vocabulary very few words are strictly *univalent* in this sense. The numerals are. In striking contrast to *we* and *you*, the personal pronoun *I* is also univalent. When one searches for other examples, however, one's choice turns almost inevitably to technical terms which have lately made their way into common speech.

The reservation implied by *lately* in the last sentence is intentional. Assimilation of technical terms in everyday speech – especially by mass media – exposes them to the process of semantic erosion responsible for the multiplicity of meaning conveyed by other words in daily use. A familiar example demonstrates the debasement of the verbal currency of science in this way. The term *allergy* came into medical use to describe a category of antigen–antibody reactions. By metaphorical extension of its meaning in a non-technical context, it now signifies to most users little more than personal dislike. *He is allergic to gendarmes* sounds more highbrow than *he has no use for cops*. Another recent corruption of this sort is the use of *dimension*. One piously hopes that the British radio parson had something meaningful to divulge in his provocative May announcement: *the Day of Pentecost added a new dimension to the hopes of despairing and disillusioned mankind.*

Medical terms are particularly liable to relinquish their proper meaning, especially if they enlarge the vernacular vocabulary of abuse. As used by people with no scientific training and often by others who should know better, *sadism* and *masochism* have entirely forfeited their definitively erotic component. In short, anyone who is callous is a sadist and anyone who seeks the martyr's crown is a masochist. Within a few years after the introduction of the word to describe hallucinatory drugs evoking very vivid sensations, the glossy magazines had begun to talk of brightly coloured wall-paper or cushion covers as *psychedelic*.[1] Inevitably, maltreatment by metonymy penetrates the domain of professional use. Though medical dictionaries draw a clear-cut distinction between *signs* (observable characteristics of a disease) and *symptoms* (the patient's subjective description of the complaint), one frequently hears medical practitioners use the word *symptom* to signify the former as well as the latter.

Unless early education can find a niche for the study of etymology with special reference to internationally current roots, another danger besets the healthy growth of the vocabulary of science. Invention by journalists and commercial firms of words based on false analogy has set a new pattern for counterfeit coinage. One example is *motorcade* by analogy with *cavalcade*, presumably on the understanding that *-cade* signifies a procession. Actually *-cade* occurs as a terminal in *cascade*, *cavalcade*, etc., only because the stem of the ancestral Latin word ended in *c*. The correct dissection of *cavalcade* is *cavalc* + *-ade* (Latin *-ata*), as in *tirade* or *accolade*.

Unhappily, it is possible to cite similar malapropisms among scientific terms introduced during the last half century. One such is *vitamin*, initially spelt *vitamine* = *vit-* (Latin *vita*) + *amine*. This combination suggests an amine essential to life. No vitamin is, in fact, an amine *sensu stricto*. The molecule of several of them (A, C, D and E) contains neither the radicle NH_2 nor even a nitrogen atom. Their designation is an even more idiotic concoction than the use of *auto-electric* by garages which stock electrical equipment for automobiles.

On all fours with the irrelevant *C* in *motorcade* is the irrelevant *R* in *positron*. *Mesons* and *photons* conform to the pattern of *electron* (ἠλεκτρον = *amber*), whose terminal is that of the nominative singular case form of a Greek neuter noun. *Neutro* is a Latin adverb meaning *in neither direction*, correctly suggesting that an alpha particle neither attracts nor repels a *neutron*. The corresponding Latin root in *positron* is *posit-* (as in *positive*). The introduction by Tukey of *bits* for binary digits has nothing but irresponsible vulgarity to commend it.

The nadir of verbal vulgarity in natural science is the designation of a unit known as the *barn*. The International Joint Commission on Standards, Units and Constants of Radioactivity sanctioned its use in 1950 because of its widespread use among physicists in the U.S.A. The barn is defined as 10^{-24} cm². It seems that it first came into use during the Manhattan District Project of World War II to signify the

[1] From ψυχη (*mind*) and δηλος fem. δηλη *manifest*, *clear*, hence *mind-clarifying*.

cross-section for interaction of slow neutrons with certain atomic nuclei. Seemingly, its derivation is referable to the American colloquialism *big as a barn*.

As the impact of scientific discoveries on our daily lives intensifies, their effect on our speech habits threatens to rob us of a benefit uppermost among the aims of Lavoisier and his associates. Without intelligent precautions to arrest a process of semantic degradation incidental to vernacular usage, the world-wide vocabulary of Western science will more and more relinquish its unique prerogatives of precision and intelligibility. The prospect is perhaps most menacing in the Anglo-American speech community and in countries where English is the second language of educated people.

Most people who use it as their first language are singularly complacent about its shortcomings. To be sure, they may justifiably congratulate themselves on shedding a load of useless flexional luggage, in particular grammatical gender. Admittedly also, they are more ready to decry the inconsistencies of its orthography than to acknowledge its etymological merits. What too few of them recognize as a speech defect, and too many regard as an amenity, is what makes it possible to write acceptably in English with a vocabulary of 850 words. C. K. Ogden[1] was able to stupefy intelligent people by so sophisticated a confidence trick only because a single word of vernacular English can convey a very large number of by no means manifestly related meanings.

It is credible that the first French critic of the claims of basic English to become a global auxiliary language dismissed it diagnostically with the comment: *C'est la maladie anglaise.* For the French take words seriously, and *L'Académie Française* guards their meaning jealously. So much so, that Renan could assert with legitimate pride the possibility of saying everything in the language of his well-bred compatriots without being pedantic. More than Catholic influence, this accounts for a widespread preference for French as *la langue diplomatique.* In diplomacy, war may be the penalty of double talk, and the penalty of atomic war may be extinction of human life.

Unlike the French, the Anglo-Saxon attitude is either indifference to the way in which words come to have more than one meaning with no recognizable common component or pride in the so-called *richness* of meaning which a single word can convey. Indeed, a

[1] Planner of *Basic English* which enjoyed a brief spell of notoriety when proposed for a global auxiliary, 1930–45. The British Council sealed its fate by espousing its cause.

widespread Anglo-American *mystique* dignifies speech as a creative activity and endorses the *non sequitur* that attempts to plan its proper use are therefore doomed to failure. Accordingly, the dictionary is a record of how a sufficiently large number of half-literate immigrants talk. For instance, Webster's latest edition gives its *nihil obstat et imprimatur* for substituting *have* by *of* in such shanty-town constructions as *would of, could of*. Doubtless, the *Concise Oxford Dictionary* will soon invoke the authority of the British Ministry of Health for *drinka pinta milka day*.

If one is a poet or a politician such a mental posture is intelligible. Before one can gain attention or assent, one has then to evoke interest by manipulating words with many emotive overtones and a wide range of associations with no essential relevance to the factual or logical content (if any) of the topic. Communication about their work between scientific workers is not like this. There exists a common bond of interest at the outset. The main concern of the transmitter is therefore to convey a message which the recipient can decode with minimum risk of misunderstanding. In so far as long-windedness and stodginess cause attention to flag, the writer should strive to avoid both; but the primary concern of a prose style tailored to the end in view is *precision*.

The *sine qua non* of precision is choice of what the French call the *mot juste*. Since early education does little to make English-speaking people alert to the semantic defects of their own language, it is therefore fitting to discuss what words do or do not fulfil this requirement. When precision is the primary desideratum, the *mot juste* is the least *plurivalent* word which can convey the appropriate meaning. Nearly all words in common use are plurivalent (i.e. have more than one meaning); but some are much more so than others. Though we may speak of the *Founding Fathers*, a *Mother Superior*, the *Brotherhood of Man* and a *Nursing Sister*, the words *father, mother, brother* and *sister* have a unique meaning in a *live context*. If qualified as in the foregoing examples, their meaning is usually unique.

What words are or are not good words in this sense should be clear, if we contrast the use of *father* with that of *factor*. Factor has at least three accredited meanings, by no means clearly related but each distinguishable in a specific context:

(i) in the context of the Scottish countryside it means a land agent;
(ii) in the context of computation or mathematics it means a divisor which leaves no remainder;

(iii) in the context of genetical discussion, when preceded by *multiple*, it means one of a group of genes which accumulatively determine a quantitative characteristic.

In its appropriate milieu, each of the foregoing three satisfies the criterion of the *mot juste*; but there is no conceivable justification for the all too common use of *factor* when the speaker or writer at a loss for words intends to convey either *component, circumstance, consideration, contributory cause* or *aspect*. It would then be more honest to substitute *what-not*.

When the meaning of a more familiar and native word tallies with that of a less familiar word of Latin origin, recourse to the latter conveys a bogus claim to precision. Thus the use of *commencement* for *beginning* has nothing to commend it. Nor has *currently* for *now* or *then*. None the less, the plea for use of *plain* words can too easily become a permit for ambiguity. By plain words, one usually means words in most frequent use; and words in most frequent use are words most vulnerable to misuse through ignorance and diversification of meaning for rhetorical effect. It is a salutary exercise in semantics to construct sentences to illustrate the different meanings of each of the following thirty-five English words, all of which rank very high in tables of word frequency.

ALWAYS: 1. on each occasion; 2. ceaselessly.

ANY: 1. even a little; 2. even one; 3. even a few; 4. every.

APPARENT: 1. seeming; 2. manifest.

AREA: 1. surface metric; 2. district; 3. territory; 4. domain; 5. sector; 6. department; 7. aspect; 8. context; 9. part; 10. *milieu*; 11. region.

AS: 1. in the way that; 2. in the like measure to [*as or so . . . as*]; 3. while; 4. because.

BAR: 1. rod; 2. drinking place; 3. prisoner's place in court; 4. advocate's profession; 5. obstacle; 6. except; 7. prevent

BRIGHT: 1. shining; 2. conspicuous; 3. intense; 4. intelligent.

CLEAR: 1. manifest; 2. understandable; 3. translucent; 4. cloudless.

DULL: 1. not shining; 2. blunt; 3. unintelligent; 4. wearying or unstimulating; 5. cloudy.

EVER: 1. even once; 2. at all times.

FAIR: 1. pale; 2. equitable; 3. beautiful; 4. neither good nor bad; 5. gathering for sale of goods.

FINE: 1. very small and/or thin; 2. dry, cloudless, sunny; 3. good, fitting, worthy of esteem; 4. financial penalty.

FOR: 1. on behalf of; 2. as a means of; 3. with a view to; 4. instead of; 5. because of; 6. in favour of.

FUNNY: 1. comic; 2. unusual; 3. impertinent; 4. alarming.

GENERALLY: 1. universally; 2. often.

HIGH: 1. tall; 2. exalted; 3. putrefying; 4. shrill; 5. drunk.

JUST: 1. equitable; 2. very recently; 3. wholly; 4. good; 5. with difficulty.

LAST: 1. final; 2. preceding; 3. [as verb] continue; 4. foot-mould.

LEFT: 1. sinistral; 2. residual; 3. departed; 4. abandoned.

ORDER: 1. rank in a sequence; 2. tidiness; 3. readiness; 4. command; 5. reservation; 6. request; 7. religious denomination; 8. secular organization; 9. official decoration.

ONLY: 1. one; 2. sole; 3. no more than; 4. nothing except; 5. without a brother or a sister.

PRACTICALLY: 1. empirically; 2. almost; 3. competently.

PRETTY: 1. pleasing; 2. somewhat; 3. very; 4. almost.

QUITE: 1. somewhat; 2. wholly; 3. I agree.

RATHER: 1. somewhat; 2. preferably.

RIGHT: 1. dextral; 2. correct; 3. privilege; 4. 90°.

SENSIBLE: 1. detectable by the sense organs; 2. equipped with sense organs; 3. intelligent; 4. judicious.

STRONG: 1. [physically] powerful; 2. not prone to bad health; 3. resolute; 4. intense; 5. concentrated; 6. not easily breakable, conquered or disposed of; 7. emphatic; 8. intoxicating; 9. safe; 10. with many allies.

SO: 1. that or it; 2. as; 3. very; 4. in this way; 5. true; 6. SO THAT = in order that; 7. SO THAT = with the result that; 8. SO Q THAT = Q to such an extent that.

SOME: 1. one sort of; 2. several; 3. even a few; 4. even a little.

STILL: 1. motionless; 2. soundless; 3. till now or till then; 4. despite that; 5. distilling apparatus.

TOO: 1. also; 2. excessively.

TRY: 1. attempt; 2. judge, arraign; 3. vex; 4. test.

WEAK: 1. [physically] feeble; 2. sickly; 3. irresolute; 4. easily breakable, conquered, destroyed or disposed of; 5. dilute; 6. unsafe; 7. with few allies.

WITH: 1. in the company of; 2. by means of.

*   *   *

To one among several ways in which we come to spell and to write in the same way words with totally different meanings, Anglo-American usage is especially prone. The reason is the diversity of sources from which English has accumulated its present vocabulary. Besides its substratum of Teutonic words essential to intelligible communication, it has acquired a considerable infusion of corrupt Latin through the Norman Conquest and Plantagenet campaigns, of classical Latin through church, law and scientific scholarship, a by no means trivial equipment based on Greek roots, a small battery of Moorish and Persian ingredients through Moslem science and the Crusades, together with a miscellany of Indian, Chinese and African words incidental to trade and colonization. English words now spelt and pronounced alike may therefore have no common ancestor.

In the glossaries of Parts II and III, we have met with one conspicuous example of this. It is of interest ulterior to our theme because it had homicidal consequences in the domain of Christian controversy. We meet the nominative singular case form of Latin HOMO in *homicide* and (as more usually) the full stem of other case forms in *hominoid*, *Hominidae*. Greek contributes to words such as *homology*, *homosexual*, *homogeneity* the root *homo-* referable to either of two cognate adjectives:

> ὅμοιος, ὅμοια, ὅμοιον, *etc.* meaning *similar, alike.*
> ὅμος, ὅμη, ὅμον, *etc.* meaning *same, identical.*

A major theological difference of opinion between the victorious Aryan Goths and their Imperial Athanasian opponents involved whether copyists of Holy Writ had correctly transcribed the Son as of *like* (ὅμοια) substance to, or as being of the *same* (ὅμη) substance as, the Father. Uncertainty about the authenticity of a single word had sanguinary consequences to believers of both persuasions.

Without recourse to a dictionary, e.g. the *Concise Oxford Dictionary*, which cites etymological sources, it is thus difficult to distinguish between two sorts of words spelt and pronounced alike. Some of which have widely different meanings may have a common ancestor. Others (so-called *homonyms*) have not. Thus it will not be so obvious to most of us as to the late C. K. Ogden that a love *match*, a cricket *match* and a good *match* of colours have much more in common than the single *match* with which a good Boy Scout is able to light a camp fire. The ancestor of the first three is the Old English word *gemaeca*, cognate with the modern word *make*. The ancestor of the Boy Scout *match* is an Old French word (now written *mèche*) for a *wick*.

British readers who are amateur or even professional botanists may have asked themselves: what is the connexion between the solitary species of *Lychnis* called *corn cockle* in the countryside with the clam called *cockle*. The answer is none. *Coccul* or *coccyl* is the nominative singular form of a word used in the Anglo-Saxon gospel version of the parable of the sower to translate what the 1611 Bible calls *tares*. It may be a diminutive remotely related to Latin (*coccus*) for *berry*. The vernacular name of the clam comes from French *coquille* for a shell. (Latin *conchylia*, from Greek κογχυλιον, diminutive of κογχη, for *mussel*.)

One way in which one can become more sensitive to the claims of precision is to allow one's knowledge of a foreign language to throw light on the semantic inadequacies of one's own. Indeed, defects of our own language may not occur to us unless the study of another language brings them into focus. Welsh clarifies the meaning of three words in the foregoing list.

The word *with* in Anglo-American, *met* in Dutch, *mit* in German, *med* in Scandinavian, *avec* in French, *con* in Spanish and Italian, covers two different notions distinguished in Welsh as follows:

> GYDA = *in the company of*, e.g. *came with her boy*
> AG = *by means of*, e.g. *cut with a knife*

Welsh also makes a significant distinction between two meanings of Anglo-American *always*, Dutch *altijd*, German *immer*, Scandinavian *alltid* or *altid*, French *toujours*, Italian *sempre*, Spanish *siempre*. No one Welsh word covers the foregoing. Instead we distinguish:

> YN WASTAD = *ceaselessly*, e.g. *always* love you, honey
> BOB AMSER = *on each occasion*, e.g. *always* brought his flute

Also from Welsh, notice the following exhibit for English *last*:

> OLAF = *final* (as in *last judgement*)
> DIWEDDAF = *preceding* (as in *last Christmas*)
> PARHAU = *persist, endure*

All languages rely to some extent on context to clarify semantic distinctions between different meanings the same word may have. In English there are four ways in which context can provide a protective umbrella for the *mot juste*:

(*a*) a qualifier may suffice, as when one speaks of *multiple factors* or *foster mothers*;

(*b*) in juxtaposition, two or more plurivalent words may convey a unique meaning, e.g. *just as* = *in the manner that*;

(*c*) the several meanings a word can have may occur in domains of discourse which do not commonly overlap, e.g. *left* (=abandoned);

(*d*) location of a word in the sentence landscape may restrict what it means, e.g.:

(i) *So* he did not (so = *therefore*)
(ii) It was not *so* (so = *true*)
(iii) Not *so* many as (so = *as*)

The most important example of (*d*) is the use of *only*. Like its equivalent in German (*nur*) and Swedish (*bara*), this word has no single equivalent in French; but if used with circumspection by its position can convey its appropriate meaning, as the following example illustrates:

ONLY the bishop gave the baboon the bun
The ONLY bishop gave the baboon the bun
The bishop ONLY gave the baboon the bun
The bishop gave ONLY the baboon the bun
The bishop gave the ONLY baboon the bun
The bishop gave the baboon ONLY the bun
The bishop gave the baboon the ONLY bun

When a word has several meanings, a writer who aims at a high level of precision will (as far as possible): (*a*) reject it if it has no indispensable use; (*b*) otherwise, use it only with its least dispensable meaning or with due regard to the safeguards of context. Lack of due respect to contextual safeguards is especially objectionable when a word has a highly specific meaning in the domain of scientific discourse. Three such words occur in the foregoing list: AREA; GENERALLY; SENSIBLE. The second of these has a unique and indispensible significance in mathematics and logic. Thus a *general* proof is a proof that a proposition is *generally*, i.e. in all circumstances, true. Since meteorology is a highly mathematical (though by no means *exact*) science, it is deplorable to hear (as frequently on the British radio weather forecast): '*it will be generally dry but with short showers in a few places.*' It would be more precise and equally informative to say: 'except in a few places which will have short showers, it will be dry.'

Contextual safeguards as an instrument of precision have their own pitfalls. One of these is attachment of a qualifying phrase which makes nonsense. It is one thing to recognize that a unique association with others may confer on a particular word an unmistakable meaning different from its connotation in its own right, e.g. a *point of view* in contradistinction to the *point at which AB intersects CD*. It is another thing to condone such ridiculous combinations as *centre around* and *to a degree*. It is unexceptionable both to *orbit around* meaning one thing and to *centre on* meaning another. *To a degree* means nothing at all unless qualified by such adjectives as *high* or *low* to indicate which end of a metaphorical scale is the writer's or speaker's intention.

A pessimistic attitude to the possibility of controlling the growth of the vocabulary of modern science, to conserve the gains of the past and to assure the same benefits as its scope enlarges, is intelligible only if we are unable to distinguish clearly between different ways in which the vocabulary of a language can change. Aside from slipshod habits of daily speech we may speak of three processes at work as *rhetorical*, *taxonomical* and *analytical*.

(i) *Rhetorical* here signifies use of those figures of speech employed to confer entertainment value, approval or antipathy on what would otherwise be a bare statement of fact or opinion. At one time it was part of a liberal education to be able to name them,[1] though not to recognize the diversification of meaning they impose on individual words by frequent use. One self-adjusting process called into action to compensate for resulting damage to the lines of communication is the creation of *idioms*, i.e. sequences (e.g. *put up with* = tolerate) with a unique meaning not deducible from that of the component words.

(ii) *Taxonomical* (i.e. classificatory) is a self-explanatory term for the process of creating new names for new objects, activities or concepts by intelligible borrowing or compounding and by unintelligible ingenuity (e.g. *nylon*).

(iii) *Analytical* here signifies the possibility of dispensing with some words by their breakdown into components, one or more of which may call for creation of a new one. A very primitive speech community may have separate words for black cows, white cows and red cows but no word for a *cow* in general. Of itself, the adoption of the four words *black*, *white*, *red* and *cow* increases the stock-in-trade of words, but it dispenses with the need for many others. A merit of English is that it is well equipped with classificatory terms such as *containers*, *fasteners* and *vehicles* which simplify definition and dispense with a multiplicity of more specialized words. Thus the single word *boat* embraces about forty names ranging from coracles, canoes, catamarans and cruisers to dinghies, liners, punts, schooners and yachts. Either the context or a qualifier suffices to specify which of them, if amusing or informative.

Of the three foregoing processes which contribute to the way in which the vocabulary of a language changes, the last two are *en rapport* with the precision science demands of its devotees. When it invades the territory of terminology, the first is wholly harmful to its healthy growth. Through the intrusion of scientific nomenclature into daily speech, this is a danger we can forestall only by recognizing it as such. If too few of us care one way or the other, it is because too few of us have made the comparatively small effort to acquaint ourselves with the basic ingredients which go to the making of what is now a world-wide vocabulary. Because too few of us have had either the opportunity or the inclination to do so, another danger lies ahead. Too

[1] Simile, metaphor, metonymy, synecdoche, transferred epithet, litotes, hyperbole, oxymoron, personification, puns.

few of us appreciate that there are intelligible and worthwhile rules of word-building. Consequently, an invasion by words whose build-up has no intelligible rationale threatens the past and future vocabulary of science. An outcome so deplorable is happily avoidable, but only if future education of scientific workers finds room for a nodding acquaintance with how the world-wide vocabulary of Western science has come into being, what advantages we owe to it and what benefits it can confer on later generations.

Appendix

Here follow separately in alphabetical order some of the most common Latin and Greek roots found in technical terms. The grammatical conventions are as cited at the beginning of Chapter 6. The numbers in parentheses after each item of the English column refer to the classifications of Parts II and III, e.g. (5) *The Four Elements and Related Words* (p. 68).

ALPHABETICAL LIST OF LATIN WORDS

a	away from (1)	agro	field (9)
ab	away from (1)	ala	wing (24)
abdomen	abdomen (24)	albo	white (4)
abies	fir (26)	alga	seaweed (26)
acer	bitter (18)	algidus	cold (18)
acerbus	acid, sharp (18)	alieno	debt (15)
aceto	vinegar (11)	alimento	food (11)
acidus	acid, sharp (18)	alius	other (18)
acie	battle array (15)	allio	garlic (26)
		altare	altar (13B)
actione	action (17)	alter	one or other of two (18)
actor	actor (15)		
acu	needle (14)	altus	high or deep (18)
acumine	point (3)	alumine	alum (10)
acutus	sharp (18)	ambo	both (2)
adipe	fat (10)	ambo	both (18)
administer	servant (15)	ambulo	I walk (20)
administrator	steward (15)	amethysto	amethyst (10)
adolescentia	youth (17)	amo	I love (20)
adoratione	worship (17)	amore	love (17)
adversus	opposite (18)	anas	duck (25)
aedificatore	builder (15)	ancilla	slave (15)
aedificio	building (13A)	ancora	anchor (14)
aegrotus	ill, sick (18)	anguilla	eel (25)
aequus	equal (18)	anguis	snake (25)
aestate	summer (6)	angulo	angle (3)
aestus	current (9)	angustus	narrow (18)
aethere	air (5)	anima	breath (5)
aethere	sky (8)	anima	breath (24)
agnus	lamb (25)	animal	animal (25)
agricola	farmer (15)	animus	mind (17)

anno	year (6)	avena	oats (26)
an[n]ulo	ring (3)	avis	bird (25)
annulus	ring, signet (12B)	axilla	armpit (24)
anser	goose (25)	axis	axle (14)
ante	before (1)	bacca	berry (26)
antiquus	old (18)	balaena	whale (25)
antrum	cave (5)	balteo	belt, girdle (12B)
apis	bee (25)	barba	beard (24)
aqua	water (5)	barbarus	foreign (18)
aquila	eagle (25)	base	base (3)
aquilone	north wind (7)	basilica	church (13A)
aranea	spider (25)	basilica	palace (13A)
aratro	plough (14)	bello	war (17)
arbitrio	verdict,	bene	well (18)
	judgement (15)	bibliotheca	study (13B)
arbor	tree (26)	blatta	cockroach (25)
arca	box (12A)	bombyx	silk (10)
arcu	arch (13B)	bono	good (19)
arcu	bow (14)	bonus	good (18)
arena	sand (5), theatre	bos, bovis	ox (25)
	(13)	brassica	cabbage (26)
areo	I am dry (20)	brevis	short (18)
argento	silver (10)	bubo, -onis	owl (25)
argilla	clay (5)	bucca	cheek (24)
armatura	armour (12B)	bufo, -onis	toad (25)
aroma	spice, seasoning	bulla	bubble (5)
	(11)	butyro	butter (11)
arte	art (16)	caballus	horse (25)
articulus	joint (24)	caecus	blind (18)
asbestos	asbestos (10)	caedo	I kill (20)
asinus	donkey (25)	caelo	sky (8)
asylum	sanctuary (13A)	caelo	heaven (8)
atrium	hall (13B)	caeruleo	blue (4)
atro	black (4)	calcaneum	heel (24)
auctumno	autumn (6)	calceo	slipper, shoe (12B)
audio	I hear (20)	calculus	pebble (5)
auditio	hearing (24)	calculus	stone (10)
augeor	I grow (20)	calice	cup (12A)
aura	air (5)	caliga	boot,
aure	ear (24)		legging (12B)
auro	gold (10)	callus	lump, callus (17)
aurora	dawn (6)	calor	heat (5)
austerus	austere (18)	calvaria	skull (24)
austro	south (9)	calx	lime (10)
auxilio	help (17)	calx	heel (24)

campus	plain (9)	*cineraceo*	grey (4)
canale	tube, pipe (14)	*cinere*	ashes (5)
cancer	crab (25)	*cingulum*	belt, girdle (12B)
candela	candle (12A)	*circulo*	circle (3)
candido	white (4)	*circum*	around (1)
candor	brightness (17)	*cirrus*	curl (24)
canis	dog (25)	*cista*	box (12A)
canistro	basket (12A)	*cisterna*	cistern (13B)
canna	reed (26)	*cithara*	lute (14)
cannabis	hemp (26)	*civi*	citizen (15)
cantu	song (17)	*civitate*	state (15)
capillo	thread (14)	*clarus*	bright (18)
capillo	hair (24)	*claustra*	bar, bolt (14)
capite	head (24)	*clava*	club (14)
capro	goat (25)	*clave, -i*	key (14)
capsa	box (12A)	*clavus*	pin (14)
captivo	prisoner (15)	*clepta*	thief (15)
carbone	coal (10)	*clivus*	hill (9)
carcere	prison (13A)	*coccyx*	cuckoo (25)
cardo	hinge (14)	*cochlea*	spiral (3)
carina	keel (14)	*cochlea*	snail (25)
carne	meat, flesh (11)	*cochleare*	spoon (12A)
caseo	cheese (11)	*cogitatione*	reflection,
castanea	chestnut (26)		contemplation
castigatione	punishment (15)		(17)
catapulta	sling, catapult (14)	*colle*	hill (9)
cataracta	waterfall (9)	*collo*	neck (24)
catena	chain (14)	*colo*	I dwell (20)
cauda	tail (24)	*colono*	farmer (15)
caulis	stem, stalk (26)	*color*	colour (4)
causa	cause (17)	*coluber*	snake (25)
caute	cliff (9)	*columba*	pigeon (25)
caverna	cave (5)	*columna*	column, pillar
cavo, cava	hollow (18)		(13B)
celer	quick (18)	*commercio*	commerce (15)
centro	centre (3)	*communis*	common (18)
centum	hundred (2)	*compactus*	compact (18)
cera	wax (10)	*(con)cavus*	hollow (18)
cerebrum	brain (24)	*conclavi*	room (13B)
cervix	neck (24)	*concussione*	shaking, shock
charta	paper (16)		(17)
cibo	food (11)	*(con)densus*	thick (18)
ciconia	stork (25)	*condimento*	spice, seasoning
cilium	eyelash (24)		(11)
cincto	belt, girdle (12B)	*conditione*	proposition (17)

conformatione	formation (17)	cultro	knife (12A)
consensus	agreement (17)	cum	together with (1)
consideratione	reflection, contemplation (17)	cumulus	heap (17)
		cuneo	wedge (14)
		cuniculus	rabbit (25)
consuetudine	custom (15)	cupro	copper (10)
(con)tactu	touch, sense of (24)	cura	care (17)
		curator	overseer (15)
contentione	contest (17)	curriculum	race, running (17)
continens (terra)	continent (9)	curru	chariot (14)
contra	against (1)	cursu	race, running (17)
controversia	debate, argument (16)	curvus	curved (18)
		cuspide	point (3)
convivio	feast (11)	cutis	skin, hide (24)
copula	bond (17)	cylindro	cylinder (3)
coquo	I cook (20)	cyprinus	carp (25)
corbe	basket (12A)	daucus	carrot (26)
corde	heart (24)	de	down (1)
corium	layer (17)	debito	debt (15)
corium	skin, hide (24)	decem	ten (2)
cornu	horn (24)	defectione	eclipse (8)
corona	wreath (12B)	deliberatione	reflection, contemplation (17)
corpus	body (24)		
cortex, -icis	bark (26)		
costa	rib (24)	dente	tooth (24)
coxa	hip (24)	deo	god (17)
credo	I believe (20)	desiderium	longing, desire (17)
cremo	I burn (20)		
creo	I create (20)	devoro	I devour (20)
crepusculo	twilight (6)	dexter	right (18)
cresco	I grow (20)	diabolo	devil (15)
creta	chalk (5)	diademate	crown (12B)
cribro	sieve (14)	dico	I say, speak (20)
crine, -i	hair (24)	dictator	dictator (15)
cruor	blood (24)	die	day (6)
crus	leg (24)	digito	finger, toe (24)
crystallo	crystal (10)	diluvium	flood (9)
cubiculo	bedchamber (13B)	disciplina	training (16)
cubiculo	room (13B)	dispositione	arrangement (17)
cubo	cube (3)	distans	distant (18)
cuculus	cuckoo (25)	distribuo	I distribute, divide (20)
culex	gnat (25)		
culmine	height (17)	diversus	different (18)
culmo	straw (10)	divido	I distribute, divide (20)
cultello	peg, stake (14)		

doceo	I teach (20)	*fabula*	fable (16)
doctor	teacher (15)	*facilis*	easy (18)
dolabra	axe (14)	*faeces*	refuse (24)
domino	despot (15)	*fagus*	beech (26)
domino	master (15)	*fallo*	I deceive (20)
domo	house (13A)	*falx*	sickle, scythe (14)
dono	gift (15)	*fame*	hunger (11)
dono	gift (17)	*fano*	temple (13A)
dormio	I sleep (20)	*farina*	flour (11)
dormitorio	room (13B)	*farina*	meal (11)
dorso	back (24)	*fastigium*	height (17)
duce	commander (15)	*fatuus*	foolish (18)
dulcis	sweet (18)	*fauces*	throat (24)
duo	two (2)	*febre, -i*	fever (24)
duodecim	twelve (2)	*felis*	cat (25)
duplex	double (18)	*femina*	woman (15)
durus	hard (18)	*femur*	thigh (24)
e	out of (1)	*fenestra*	window (13B)
ecclesia	church (13A)	*fera*	animal (25)
effigie	image (12A)	*fermento*	yeast, leaven (11)
ego	I (1)	*fero*	I carry (20)
electro	amber (10)	*ferro*	iron (10)
elegans	elegant (18)	*ferula*	rod (14)
emitto	I shoot (20)	*ferus*	wild (18)
ense	sword (14)	*fervens*	hot (18)
episcopo	bishop, overseer (15)	*fibra*	fibre (10)
epistola	letter (16)	*ficus*	fig (26)
equus	horse (25)	*filio*	son (15)
errone	wanderer (15)	*filo*	thread (14)
eruditione	learning (16)	*findo*	I split (20)
esca	food (11)	*fine*	end, purpose (17)
euro	east wind (7)	*fisso*	cleft (5)
ex	out of (1)	*fistula*	tube, pipe (14)
exactio	tax, tribute (17)	*fistula*	whistle (14)
excrementa	excrement (24)	*flagello*	I whip (20)
exemplum	model, type (17)	*flagellum*	whip (14)
exiguus	scanty (18)	*flamma*	flame (5)
experientia	trial (17)	*flatu*	wind (5)
exteriori	exterior	*flecto*	I bend (20)
extra	outside (1)	*flore*	flower (26)
extra	outside (19)	*florens*	flowering (18)
extremo	extreme	*fluctu*	wave (5)
faba	bean (26)	*fluento*	stream (9)
fabro	artisan (15)	*flumine*	river (9)
		fluo	I flow (20)

fluvio	river (9)	*genu*	knee (24)
f(o)eno	hay (10)	*genus*	tribe, clan (15)
folio	leaf (26)	*giganteus*	gigantic (18)
folle	bellows (14)	*gingiva*	gum (24)
folliculo	bag (12A)	*glacie*	ice (5)
fonte	fountain, spring (5)	*gladius*	sword (14)
foramen	hole, cavity (17)	*glauco*	green (4)
forceps	vice (14)	*globo*	sphere (3)
formica	ant (25)	*glutine*	glue (10)
fortuna	riches (17)	*gradior*	I walk (20)
forum	market (15)	*gradu*	step (13B)
fossa	ditch (13B)	*gramine*	grass (26)
fovea	pit, hole (9)	*granariis*	barn (13A)
frango	I break (20)	*grandine*	hail (7)
fratre	brother (15)	*grandis*	big (18)
fraxinus	ash (26)	*gravis*	heavy (18)
frenum	bridle (14)	*grege*	herd, flock (17)
freto	strait (9)	*grossus*	thick (18)
frico	I rub (20)	*gryllus*	grasshopper (25)
frigidus	cold (18)	*gubernaculum*	rudder (14)
fronde	leaf (26)	*gubernatore*	pilot (15)
fronte	forehead (24)	*gula*	throat (24)
fructu	fruit (26)	*gurgite*	whirlpool (9)
fruge	fruit (26)	*gutta*	drop (5)
fulgeo	I shine (20)	*habito*	I dwell (20)
fuligine	soot (5)	*halitus*	breath (5)
fulmine	lightning (7)	*hallex*	big toe (24)
fumo	smoke (5)	*hamo*	hook (14)
fune	cord (14)	*hasta*	spear (14)
funere	funeral (15)	*hebdomade*	week (6)
fungus	mushroom (26)	*hebes*	blunt (18)
furca	fork (12A)	*herba*	herb (26)
fure	thief (15)	*hiatus*	cleft (5)
furno	oven (13B)	*hieme*	winter (6)
galea	helmet (12B)	*hirsutus*	hairy, rough (18)
gallina	hen (25)	*hirudo*	leech (25)
gallus	(domestic) fowl (25)	*hirundo*	swallow (15)
		histrione	actor (15)
gelidus	cold (18)	*(h)olere*	vegetable (26)
gemma	bud (26)	*homicida*	killer (15)
gena	cheek (24)	*homo*	human being (15)
generalis	general, universal (18)	*hora*	hour (6)
		hordeum	barley (26)
genere	birth (17)	*hortu*	garden (13B)
genero	I create (20)	*hospite*	host (15)

hospite	stranger, foreigner (15)	*interiore*	inner (19)
hospitio	tavern, inn (13A)	*interprete*	interpreter (15
hoste	enemy (15)	*intimo*	innermost (19)
humano	man, mankind (15)	*intra*	inside (1)
		intra	inside (19)
		invenio	I discover (20)
humerus	shoulder (24)	*involucrum*	covering (12B)
humidus	moist, wet (18)	*ipse*	self (1)
humus	ground, soil (5)	*iride*	rainbow (7)
hydrargyrum	mercury (10)	*isthmo*	isthmus (9)
hymno	hymn (16)	*iterum*	again (18)
lacio	I throw (20)	*itinere*	road (9)
igni	fire (5)	*iudice*	judge (15)
im-	in (1)	*iudicio*	trial (17)
imagine	image (12A)	*iugum*	yoke (14)
imbre	shower (7)	*iuniore*	younger (19)
imbrex	tile (13B)	*iure*	law (15)
imitatione	imitation (17)	*ius*	broth (11)
immortalis	immortal (18)	*iuvene*	young (19)
impar	odd, uneven (18)	*labium*	lip (24)
imperatore	commander (15)	*labore*	work (17)
imperio	state (15)	*laboro*	I work (20)
impero	I govern (20)	*labrum*	lip (24)
in	in (1)	*lacerta*	lizard (25)
incendo	I burn (20)	*lacrima*	tear (24)
inceptione	beginning (17)	*lacte*	milk (11)
inchoatus	imperfect (18)	*lactuca*	lettuce (26)
infante	child, boy (15)	*lacu*	lake, pool (9)
inferiore	lower (19)	*lacuna*	lake, pool (9)
infimo	lowest (19)	*lacuna*	hole, cavity (17)
infra	below (19)	*laevus*	left (18)
infundibulo	funnel (14)	*lamentatione*	lament (17)
ingens	gigantic (18)	*lamina*	layer (17)
inimico	enemy (15)	*lamina*	scale (24)
initio	beginning (17)	*lana*	wool (10)
innumerabilis	innumerable (18)	*lana*	wool (24)
insania	madness (24)	*lanatus*	woolly (18)
instituto	custom (15)	*lancea*	spear (14)
instrumento	device (14)	*laniger*	woolly (18)
instrumento	tool (14)	*lapide*	tombstone (14)
insula	island (9)	*lapis*	rock (5)
integer	whole (18)	*lapis*	stone (10)
integumento	veil (12B)	*latere*	side (24)
intensione	stretching (17)	*latus*	broad (18)
inter	between (1)	*legato*	ambassador (15)

lege	law (15)	*mala*	cheek (24)
leo, -onis	lion (25)	*malleus*	hammer (14)
lepus	hare (25)	*malo*	mast (14)
letho	death (17)	*malo*	apple (26)
liber	free (18)	*malo*	bad (19)
liberatione	release (17)	*malus*	bad (18)
libertate	freedom, liberty (17)	*mamma*	breast (24)
		mancipio	slave (15)
libido	longing, desire (17)	*mandibula*	jaw (24)
		manifestus	manifest (18)
libro	book (16)	*manu*	hand (24)
ligno	wood, timber (10)	*mare*	sea (9)
		margarita	pearl (10)
lima	file (14)	*marmore*	marble (10)
limen	threshold (13B)	*massa*	mass, lump (17)
limo	mud (5)	*matre*	mother (15)
linea	line (3)	*matutino*	morning (6)
lingua	tongue (24)	*maxilla*	jaw (24)
linum	flax, linen (10)	*maximo*	maximal (19)
litore	shore (9)	*maximus*	most (18)
littera	letter (ABC) (16)	*medicamento*	drug (24)
locus	place (17)	*medicina*	medicine (24)
locusta	locust (25)	*medico*	physician (15)
locutione	phrase (17)	*medico*	medical (24)
longe	afar (18)	*medicus*	medical (18)
longus	long (18)	*medio*	centre (3)
lorica	breastplate (14)	*medius*	middle (18)
lorum	bridle (14)	*meliore*	better (19)
lubricus	smooth (18)	*melle*	honey (11)
luce	light (17)	*memoria*	memory (17)
luceo	I shine (20)	*mensa*	table (12A)
lucerna	lamp (12A)	*mense*	month (6)
luco	grove (9)	*mente*	understanding (16)
lumen	light (17)	*mente*	mind (17)
luna	moon (8)	*mergo*	I dip (20)
lupus	wolf (25)	*meridie*	south (9)
luteo	yellow (4)	*meridie*	noon (6)
lyra	lyre (14)	*metallo*	mine (5)
macer	lean (18)	*metior*	I measure (20)
machina	machine (14)	*milite*	soldier (15)
magistro	teacher (15)	*mille*	thousand (2)
magno	big (19)	*minimo*	minimal (19)
magnus	big (18)	*minister*	attendant (15)
magus	magician (15)	*minister*	servant (15)
maiori	bigger (19)	*minore*	minor (19)

164

misceo	I mix (20)	nece	death (17)
misericordia	pity (17)	nemore	forest, wood (9)
missile	missile (14)	neo	I spin, whirl (20)
mixtura	mixture (17)	nervo	nerve (24)
mobile	moveable (18)	nigro	black (4)
moderator	ruler (15)	nimbus	cloud (8)
mollis	soft (18)	nitro	saltpetre (10)
monade	unit (17)	nive	snow (7)
moneta	money, wealth (15)	nocte	night (6)
		nocuus	harmful (18)
monstro	I show (20)	nodus	knot (14)
monte	mountain (9)	nomine	name (17)
morbo	disease (24)	noto	south wind (7)
mordeo	I bite (20)	novem	nine (2)
mores	custom (15)	novus	new (18)
morte	death (17)	nube	cloud (8)
mortuus	dead (18)	nubo	I marry (20)
morus	mulberry (26)	nudus	naked (18)
moveo	I move (20)	numero	number (2)
mucrone	point (3)	numero	number (17)
mucrone	sword (14)	numisma	coin (15)
muliere	woman (15)	nummo	coin (15)
multo	much (19)	nuntio	messenger (15)
multi	many, much (18)	nupta	bride (15)
mundo	universe, world (8)	obliquus	crosswise (18)
		obscuritate	darkness (17)
munere	gift (15)	obtusus	blunt (18)
muro	wall (13B)	occidente	west (9)
mus, muris	mouse (25)	occulo	I hide (20)
musca	fly (25)	oceano	ocean (9)
musculo	muscle (24)	ocellus	eye (24)
museum	study (13B)	ocrea	boot, legging (12B)
musica	music (17)		
mustela	weasel (25)	octo	eight (2)
napo	turnip (26)	oculo	eye (24)
narratione	narrative (17)	odium	hatred (17)
nascor	I am born (20)	odore	odour (17)
naso	nose (24)	officina	shop (13A)
nato	I swim (20)	olea	olive (26)
natura	nature (17)	oleo	oil (11)
nauta	sailor (15)	omen	omen (17)
nave, -i	ship (14)	omnis	all (18)
nauta	sailor (15)	onager	donkey (25)
navigo	I sail (20)	opera	work (17)
nebula	mist (7)	operculum	lid (14)

operor	I work (20)	*pavimento*	floor (13B)
opinione	opinion (17)	*pax*	peace (17)
optimo	best (19)	*pecore*	animal (25)
optimus	best (18)	*pectine*	comb (14)
opulentia	riches (17)	*pectore*	breastbone (24)
opus	work (17)	*peculiaris*	private (18)
oratione	discourse (16)	*pecunia*	money, wealth
ordine	rank, arrangement		(15)
	(17)	*pede*	foot (24)
organo	machine (14)	*pedica*	trap, snare (14)
oriente	east (9)	*pediculus*	louse (25)
origine	origin (17)	*peiori*	worse (19)
ortu	birth (17)	*pelle*	skin, hide (24)
oryza	rice (26)	*pelvis*	basin (12A)
os, osse	bone (24)	*penna*	pen, stylus (14)
os, ore	mouth (24)	*penna* or *pinna*	feather, fin (24)
ostium	door (13B)	*per*	through (1)
ostrea	oyster (25)	*perdix*	partridge (25)
ovis	sheep (25)	*peregrino*	stranger,
ovum	egg (24)		foreigner (15)
paene	almost (18)	*peregrinus*	foreign (18)
paeninsula	peninsula (9)	*perfectus*	perfect, even
pago	village (15)		(18)
pagurus	crab (25)	*persica*	peach (26)
pala	spade (14)	*pessimo*	very bad (19)
pallium	cloak (12B)	*peto*	I seek (20)
palma	palm (24)	*pharetra*	quiver (14)
palpebra	eyebrow (24)	*picea*	fir (26)
palude	marsh (9)	*pictura*	picture (13B)
pane	bread (11)	*pila*	column, pillar
papilio	butterfly (25)		(13B)
parente	father (15)	*pileus*	cap (12B)
parente	parent (15)	*pilus*	fur (24)
pariete	wall (13B)	*pinna*	feather, fin (24)
parte	part (17)	*pinus*	pine (25)
parvo	small (19)	*pipere*	pepper (11)
parvus	little (18)	*pisces*	fish(es) (25)
passer	sparrow (25)	*pisum*	pea (26)
pastor	herdsman (15)	*placenta*	cake (11)
patella	dish (12A)	*plaga*	trap, snare (14)
patina	dish (12A)	*planta*	sole (24)
patior	I suffer (20)	*planus*	flat (18)
patre	father (15)	*plaustro*	cart (14)
paucus	few (18)	*plenus*	full (18)
pauper	poor (18)	*pluma*	feather, fin (24)

plumbo	lead (10)	pro	on behalf of, in favour of (1)
pluri			
plurimo		procul	afar (18)
pluvia	rain (5)	procurator	steward (15)
poemate	poem (16)	professor	teacher (15)
poena	punishment (15)	profundus	deep (18)
pollen	flour (11)	progenie	offspring, race (17)
pollex	thumb (24)		
pomo	fruit (26)	prolixus	long (18)
ponte	bridge (13A)	promontorio	cape (9)
pontifice	priest (15)	pronuntio	I announce (20)
ponto	sea (9)	prope	near (19)
populo	people (15)	propinquus	near (18)
populus	poplar (26)	propiori	nearer (19)
porco	pig (25)	proxime	beside, near (1)
porta	door (13B)	proximo	nearest (19)
porto	I carry (20)	pruina	frost (7)
possessor	owner (15)	prunus	plum (26)
possum	I am able (20)	puero	child, boy (15)
post	after (1)	pugna	battle (15)
post	behind (19)	pugna	battle (17)
posteriori	hinder (19)	pugnatore	warrior, hero (15)
postremo	hindmost (19)		
postremus	hindmost (18)	pulcher	beautiful (18)
potentia	power (17)	pulex	flea (25)
potestate	power (17)	pulmone	lung (24)
potione	drink (11)	pulso	pulse (24)
prae	in front (19)	pulvere	dust (10)
praeceptor	teacher (15)	pulvillus	pillow, cushion (12A)
praefatione	preface (16)		
praefecto	commander (15)	pulvinus	pillow, cushion (12A)
pr(a)emium	reward, prize (15)		
		pumice	stone (10)
praeside	overseer (15)	purpureo	purple (4)
prato	field (9)	pusillus	little (18)
pravus	crooked (18)	puteo	pit (9)
pre-	before (1)	putrefacio	I cause to rot (20)
prehendo	I grasp (20)	putresco	I putrefy, rot (20)
prehenso	I grasp (20)	putris	putrid (18)
pretio	price, cost (15)	pyramide	pyramid (3)
primo	first (2)	pyrus	pear (26)
primo		quadrangulo	square (3)
priori		quadrato	square (3)
pristis	sawfish (25)	quaero	I seek (20)
privatus	private (18)	quantitate	quantity (17)

quattuor	our (2)	*saccharo*	sugar (11)
quercus	oak (26)	*sacco*	bag (12A)
quinque	five (2)	*sacerdote*	priest (15)
radio	radius (3)	*s(a)eta*	bristle (24)
radio	shuttle (14)	*sagitta*	arrow (14)
radio solis	sunbeam (8)	*sale*	salt (10)
radix, -icis	root (26)	*salix*	willow (26)
ramo	branch (26)	*salmo, -onis*	salmon (25)
rana	frog (25)	*saltatione*	dance (16)
re	thing, object (17)	*sanctus*	holy (18)
rector	steersman (15)	*sanguine*	blood (24)
rector	ruler (15)	*sanitate*	health (24)
rectus	straight (18)	*sanus*	healthy (18)
rege	king (15)	*sapiens*	wise (18)
regia	palace (13A)	*sapientia*	wisdom (17)
regno	state (15)	*satellite*	attendant (15)
regno	I govern (20)	*satis*	enough (18)
rego	I rule (20)	*saxo*	rock (5)
regula	rule, rod (16)	*saxo*	stone (10)
remedio	remedy (24)	*scala*	staircase (13B)
remo	oar (14)	*scalpro*	knife (12A)
remotus	remote (18)	*scalprum*	chisel (14)
rene	kidney (24)	*scapula*	shoulderblade (24)
repo	I creep (20)		
respublica	state (15)	*scarabeo*	beetle (25)
rete	net (14)	*schola*	school (16)
retro	backwards (1)	*scientia*	science (16)
rhetorica	rhetoric (16)	*scientia*	knowledge (17)
rhombus	rhombus (3)	*scindo*	I cut (20)
rhythmo	rhythm (17)	*scintilla*	spark (5)
rima	cleft (5)	*scopae*	broom (12A)
rore	dew (5)	*scribo*	I write (20)
roseo	pink (4)	*scripto*	writing (16)
rostro	beak (24)	*sculptor*	sculptor (15)
rota	wheel (14)	*scutum*	shield (14)
rotundus	round (18)	*sebum*	fat (10)
rubro	red (4)	*secale*	rye (26)
rufus	reddish (4)	*seco*	I cut (20)
ruga	wrinkle (24)	*secundo*	second (2)
rumen	throat (24)	*secure, -i*	axe (14)
rumino	chew (20)	*sella*	chair (12A)
rumpo	I burst (20)	*sella*	saddle (14)
runcina	plane (14)	*semen, -inis*	seed (26)
rupe	rock (5)	*semi-*	half (2)
sabulo	sand (5)	*semita*	path (13B)

sene	old man (15)	*sitio*	I thirst (20)
senectute	(old) age (17)	*socco*	slipper, shoe (12B)
senex	old (18)	*socio*	ally (15)
seni	old (19)	*sole*	sun (8)
senilis	old (18)	*solea* (of foot)	sole (24)
senio	(old) age (17)	*solea* (fish)	sole (25)
seniori	senior (19)	*solidus*	solid (18)
sententia	verdict,	*solitarius*	solitary (18)
	judgement (15)	*solum*	alone (18)
sententia	opinion (17)	*somno*	sleep (17)
separo	I separate (20)	*somno*	sleep (24)
septem	seven (2)	*sono*	sound (17)
septentrione	north (9)	*sopore*	sleep (17)
septum	hedge, fence (13B)	*sopore*	sleep (24)
sepulcro	grave (13A)	*sordes*	dirt (10)
serico	silk (10)	*sorore*	sister (15)
sermone	discourse (16)	*spatha*	sword (14)
serpente	snake (25)	*spatio*	space (17)
serra	saw (14)	*spectaculo*	view, sight (17),
serro	I saw (20)		theatre (13)
servio	I serve (20)	*specto*	I look at (20)
servo	slave (15)	*speculum*	mirror (12A)
sex	six (2)	*spelunca*	cave (5)
siccus	dry (18)	*sphaera*	sphere (3)
sidere	star (8)	*spiculo*	arrow (14)
signo	sign, symbol (17)	*spina*	spine, backbone
silice	flint (10)		(24)
silva	forest, wood (9)	*spira*	spiral (3)
simia	monkey (25)	*spiritu*	breath (5)
similis	similar (18)	*spiritu*	breath (24)
similis	alike (18)	*splendeo*	I shine (20)
simplex	simple (18)	*spuma*	foam (5)
simplice	simple (2)	*spuma*	cream (11)
simulacrum	image (12A)	*squalus*	shark (25)
simulacrum	apparition (15)	*squama*	scale (24)
sinape	mustard (11)	*stamen*	thread (14)
sine	without (1)	*stanno*	tin (10)
singuli	single (2)	*statua*	statue (13B)
singuli	individual (18)	*status*	condition (17)
sinister	left (18)	*stella*	star (8)
sinu	curve (3)	*stercore*	excrement (24)
sinu	bay (9)	*sternum*	breastbone (24)
sinuosus	curved (18)	*stillo*	I drop (20)
siphone	siphon (14)	*stilus*	pen, stylus (14)
siti	thirst (11)	*stimulus*	goad, sting (14)

stipe	coin (15)	*tela*	web (12B)
stipe	gift (11)	*tellure*	earth (5)
stipendio	tax, tribute (17)	*telo*	weapon (14)
stipula	stem, stalk (26)	*tempestate*	storm (7)
stirp(s)	tribe, clan (15)	*templo*	temple (13A)
stirps	trunk (26)	*tempore*	time (6)
stragulo	carpet (12A)	*tempore*	temple (24)
stratum	blanket, covering (12A)	*tenuis*	thin (18)
		tergo	back (24)
strepitu	sound (17)	*terminus*	boundary (9)
stridor	sound (17)	*terra*	earth (5)
strigilis	comb (14)	*terra*	land (9)
strix, -igis	owl (25)	*terreo*	I frighten (20)
structore	builder (15)	*terribilis*	terrible (18)
stultus	foolish (18)	*tertio*	third (2)
suavis	sweet (18)	*tessera*	tile (13B)
sub	below, under (1)	*testa*	shell (24)
subula	awl (14)	*textura*	web (12B)
suc(c)o	juice (11)	*theatro*	theatre (13A)
succussione	shaking, shock (17)	*thorax*	breastplate (14)
		tibia	pipe (14)
sudore	sweat (24)	*timeo*	I fear (20)
sui	self (17)	*toleratione*	suffering (17)
sulfure	sulphur (10)	*tonitru*	thunder (7)
super	above, over (1)	*torque*	wreath (12B)
superiori	upper (19)	*totus*	whole (18)
supra	above, over (1)	*trabe*	beam (13B)
supra	above (19)	*tramite*	path (13B)
supremo	uppermost (19)	*trans*	across, through (1)
sus	pig (25)		
suscito	I rouse (20)	*transtro*	beam (13B)
syllaba	syllable (16)	*tres*	three (2)
tabula	table (12A)	*tribu*	tribe, clan (15)
tabula	picture (13B)	*tributo*	tax, tribute (17)
tactu	touch (17)	*tridente*	fork (12A)
taedium	weariness (17)	*triticum*	wheat (26)
talpa	mole (25)	*triumpho*	victory (15)
talus	heel (24)	*trochlea*	pulley (14)
tapete	carpet (12A)	*truncus*	trunk (26)
tardus	slow (18)	*tuba*	trumpet (14)
taurus	bull (25)	*tubo*	tube, pipe (14)
tecto	roof (13B)	*tubulo*	tube, pipe (14)
tegmen	covering (12B)	*tumulus*	hill (9)
tego	I cover (20)	*tunica*	tunic (12B)
tegula	tile (13B)	*turbine*	whirlwind (7)

tussis	cough (24)	*verbo*	word (16)
tympanum	drum (14)	*vermes*	worms (25)
ulteriore	ulterior	*verno*	spring (6)
ultimo	ultimate	*verruca*	wart (24)
ultra	beyond (1)	*versu*	row, verse (16)
ultra	beyond (19)	*vertebra*	vertebra (24)
umbo	shield (14)	*verticillo*	whirl, eddy (5)
umbra	shadow (8)	*verus*	true (18)
uncus	hook (14)	*vesica*	bladder (24)
unda	wave (5)	*vespa*	wasp (25)
ungues	claw, hoof (24)	*vespere*	evening (6)
ungula	claw, hoof (24)	*vespertilio*	bat (25)
unione	unit (17)	*vestibulo*	porch (13B)
unus	one (2)	*veste*	apparel, clothes
urbe	city (15)		(12B)
urina	urine (24)	*via*	road (9)
ursus	bear (25)	*vicinus*	near (18)
urtica	nettle (26)	*vico*	village (15)
uterus	womb (24)	*victoria*	victory (15)
utre	bag (12A)	*villa*	house (13A)
uva	bunch of grapes	*vinculum*	chain (14)
	(26)	*vindicta*	punishment (15)
vacca	cow (25)	*vinea*	vine (26)
vacuus	empty (18)	*vino*	wine (11)
vagina	sheath (14)	*viola*	violet (4)
valle	velley (9)	*vipera*	snake (25)
vapor	steam, vapour	*virga*	rod (14)
	(5)	*virgo*	virgin (15)
variatus	varied (18)	*viridi*	green (4)
varius	various (18)	*viro*	male, man (15)
vas	vessel (12A)	*viscera*	viscera (24)
vastus	big (18)	*visibilis*	visible (18)
vectis	lever (14)	*visione*	sight (24)
vehiculo	vehicle (14)	*vita*	life (17)
veho	I carry (20)	*vitis*	vine (26)
velamen	covering (12B)	*vitro*	glass (10)
velo	sail (14)	*vivo*	I live (20)
velum	curtain (12A)	*voluntate*	will (17)
vena	vein (24)	*vomere*	plough (14)
vendo	I sell (20)	*vortex*	whirlwind (7)
veneratione	worship (17)	*vortex*	whirlpool (9)
ventilo	I fan (20)	*vox*	voice (17)
vento	wind (5)	*vulnere*	wound (24)
ventre	belly (24)	*vulpes*	fox (25)
verax	true (18)	*zephyrus*	west wind (7)

ἀ-	without (1)	ἀλγος	pain (21)
ἀγαθος	good (18)	ἀλεκτρυων	cock (22)
ἀγαλματοποιος	sculptor (15)	ἀλληλων	of each other (18)
ἀγγειον	box (12A)	ἀλλος	other (18)
ἀγγελλω	I announce (20)	ἁλς, ἁλος	salt (10)
ἀγγελος	messenger (15)	ἀλσος	grove (9)
ἁγιος	holy (18)	ἀλφιτον	meal (11)
ἀγκιστρον	hook (14)	ἁμαξα	cart (14)
ἀγκος	valley (9)	ἀμβλυς	blunt (18)
ἀγκυλος	curved (18)	ἀμβροτος	immortal (18)
ἀγκυρα	anchor (14)	ἀμεθυστος	amethyst (10)
ἀγλαος	bright (18)	ἀμπελος	vine (23A)
ἀγορα	market (15)	ἀμυλος	meal (11)
ἀγριος	wild (18)	ἀμφω	both (18)
ἀγρος	field (9)	ἀν	without (1)
ἀγρωστις	grass (23A)	ἀνα	up (1)
ἀγωγη	training (16)	ἀνατολη	east (9)
ἀγων	contest (17)	ἀνεμος	wind (5)
ἀδελφη	sister (15)	ἀνεμωνη	anemone (23A)
ἀδελφος	brother (15)	ἀνηρ, ἀνδρος	male, man (15)
ἀδην	gland (21)	ἀνθεμον	flower (23B)
ἀηρ	air (5)	ἀνθηρος	flowering (18)
ἀθλον	reward, prize (15)	ἀνθος	flower (23B)
αἰγιαλος	shore (9)	ἀνθραξ	coal (10)
αἰθηρ	sky (8)	ἀνθρωπος	man, mankind (15)
αἱμα	blood (21)		
αἰολος	varied (18)	ἀνθρωπος	human being (15)
αἰσθησις	perception (17)	ἀντι	against (1)
αἰτια	cause (17)	ἀξινη	axe (14)
ἀκαληφη	nettle (23B)	ἀξων	axle (14)
ἀκανθα	spine, thorn (23B)	ἀορτη	aorta (21)
ἀκαρι	mite (22)	ἁπλοος	simple (18)
ἀκμη	point (3)	ἁπλοος	simple (2)
ἀκος	remedy (21)	ἀπο	away from (1)
ἀκουστος	audible (18)	ἀποθηκη	barn (13A)
ἀκουω	I hear (20)	ἁπτω	I grasp (20)
ἀκρα	cape (9)	ἀραιος	thin (18)
ἀκρος	high (18)	ἀραχνη	spider (22)
ἀκτη	shore (9)	ἀργυρος	silver (10)
ἀκτις, -ινος	radius (3)	ἀρθρον	joint (21)
ἀκτις	sunbeam (7)	ἀριθμος	number (17)
ἀκτις	sunbeam (8)	ἀριθμος	number (2)
ἀλαβαστριτης	alabaster (10)	ἀριστος	best (18)

ἀρκτος	bear (22)
ἀρκτος	north (9)
ἁρμα	chariot (14)
ἁρμονια	agreement (17)
ἁρπαγη	hook (14)
ἁρπη	sickle, scythe (14)
ἀρτηρια	artery (21)
ἀρτιος	perfect, even (18)
ἀρτος	bread (11)
ἀρχαιος	old (18)
ἀρχη	beginning (17)
ἀρχω	I rule (20)
ἀρχων	ruler (15)
ἀρωμα	spice, seasoning (11)
ἀσβεστος	asbestos (10)
ἀσβολος	soot (5)
ἀσκος	bag (12A)
ἀσπαραγος	asparagus (23A)
ἀστακος	lobster (22)
ἀστηρ	star (8)
ἀστραπη	lightning (7)
ἀσυλον	sanctuary (13A)
ἀτελης	imperfect (18)
ἀτμος	steam, vapour (5)
αὐλη	hall (13B)
αὐλος	pipe (14)
αὐστηρος	austere (18)
αὐτος	self (17)
αὐτος	self (1)
αὐτοχειρ	killer (15)
ἀφρος	foam (5)
ἀχατης	agate (10)
ἀψις	knot (14)
βαθυς	deep (18)
βαλλω	I throw (20)
βαπτω	I dip (20)
βαραθρον	pit (9)
βαρβαρος	stranger, foreigner (15)
βαρυς	heavy (18)
βασιλευς	king (15)
βασιλικη	palace (13A)
βασις	base (3)
βατραχος	frog (22)
βδελλα	leech (22)
βελος	missile (14)
βιβλιον	book (16)
βιος	life (17)
βιοω	I live (20)
βλαστος	bud (23B)
βλεφαρον	eyelid (21)
βομβυξ	silkworm (22)
βορβορος	dirt (10)
βορεας	north wind (7)
βοτανη	herb, plant (23B)
βοτρυς	bunch of grapes (23B)
βουκολος	herdsman (15)
βουνος	hill (9)
βους	ox (22)
βουτυρον	butter (11)
βραγχια	gills (21)
βραδυς	slow (18)
βραχιων	arm (21)
βραχυς	short (18)
βρογχος	throat (21)
βροντη	thunder (7)
βρυωνη	moss (23B)
βυσσος	calico (10)
βωμος	altar (13B)
γαλα	milk (11)
γαμεω	I marry (20)
γανος	brightness (17)
γαστηρ	belly (21)
γαστροκνημη	calf of (leg) (21)
γεινομαι	I am born (20)
γενεθλη	birthday (17)
γενεσις	origin (17)
γενετη	birth (17)
γενος	offspring, race (17)
γεννυς	cheek (21)
γεραιος	old (18)
γερων	old man (15)
γεφυρα	bridge (13A)
γεωργος	farmer (15)
γη	land (9)
γη	earth (5)
γηρας	age (old) (17)

γιγαντειος	gigantic (18)	διαβολος	devil (15)
γλαυξ	owl (22)	διαιτα	regimen (21)
γληνη	eyeball (21)	διακονεω	I serve (20)
γλυκυς	sweet (18)	διακονος	servant (15)
γλυφειον	chisel (14)	διδασκω	I teach (20)
γλυφω	I tunnel, hollow out (20)	δικελλα	fork (12A)
		δικη	trial (17)
γλωσσα	tongue (21)	δικτυον	net (14)
γναθος	jaw (21)	δινη	whirlpool (9)
γνωσις	knowledge (16, 17)	διπλοος	double (18)
γονευς	parent (15)	διπλοος	double (2)
γονη	generation (17)	διψα	thirst (11)
γραμμα	letter (ABC) (16)	διψαω	I thirst (20)
		δοκος	beam (13B)
γραμμη	line (3)	δολιχος	long (18)
γραφη	writing (16)	δοξα	opinion (17)
γραφις	pen, stylus (14)	δορυ	spear (14)
γραφω	I write (20)	δουλος	slave (15)
γυμνος	naked (18)	δραχμη	coin (15)
γυνη	woman (15)	δρεπανον	sickle, scythe (14)
γυρος	ring (3)	δρομος	race, running (17)
γυψος	chalk (5)	δροσος	dew (5)
γωνια	angle (3)	δρυς	tree (23B)
δαιω	I divide (20)	δυναμαι	I am able (20)
δακνω	I bite (20)	δυναμις	power (17)
δακτυλιος	ring, signet (12B)	δυναστης	ruler (15)
δακτυλος	finger (21)	δυο	two (2)
δασμος	tax, tribute (17)	δυσ-	bad (18)
δασυς	hairy, rough (18)	δυσις	west (9)
δεινος	terrible (18)	δωδεκα	twelve (2)
δεκα	ten (2)	δωμα	room (13B)
δενδρον	tree (23B)	δωρον	gift (17)
δεξαμενη	cistern (13B)	ἐαρ	spring (6)
δερμα	skin (21)	ἑβδομας	week (6)
δεσμος	chain (14)	ἐγκεφαλος	brain (21)
δεσμος	bond (17)	ἐγω	I (1)
δεσμωτης	prisoner (15)	ἐδαφος	floor (13B)
δεσποτης	despot (15)	ἐθνος	tribe, clan (15)
δευτερος	second (2)	εἰδωλον	image (12A)
δηλος	manifest (18)	εἰκων	statue (13B)
δημος	people (15)	εἰκων	image (12A)
δι-	two (2)	εἰρηνη	peace (17)
δια	across, through (1)	εἱρκτη	prison (13A)
		ἐκ, ἐξ	out of (1)
δια	through (1)	ἑκατον	hundred (2)

ἐκκλησια church (13A)
ἐκλειψις eclipse (8)
ἐκτομη cutting out (21)
ἐκτος outside (1)
ἐκφορα funeral (15)
ἐλαιον oil (11)
ἐλεγος lament (17)
ἐλεημοσυνη pity (17)
ἐλευθερια freedom, liberty (17)
ἐλευθερος free (18)
ἐλεφας elephant (22)
ἑλιξ tendril, spiral (23B)
ἑλιξ, -ικος spiral (3)
ἑλλεβορος hellebore (23A)
ἑλμις, -ινθος worm (22)
ἑλος marsh (9)
ἐμ- in (1)
ἐμβολος wedge (14)
ἐμβρυον embryo (21)
ἐμετος vomit (21)
ἐμπορια commerce (15)
ἐν in (1)
ἑν one (2)
ἐναντιος opposite (18)
ἐνδον inside (1)
ἐνιαυτος year (6)
ἐννεα nine (2)
ἐντερον gut (21)
ἐντομα insects (22)
ἐξ six (2)
ἑξις condition (17)
ἐπι on (1)
ἐπιβολη layer (17)
ἐπισκοπος bishop, overseer (15)
ἐπιστημη knowledge (17)
ἐπιστολη letter (16)
ἑπτα seven (2)
ἐργαζομαι I work (20)
ἐργον work (17)
ἐρεικη heath (23A)
ἐρημος solitary (18)
ἐριον wool (10)

ἐρις debate, argument (16)
ἑρπετον reptile (22)
ἐρυθρος red (4)
ἐρως love (17)
ἐσθησις apparel, clothes (12B)
ἐσοπτρον mirror (12A)
ἑσπερα evening (6)
ἐσχατος remote, furthest (18)
ἑτερος different (18)
ἐτος, ἐτεος year (6)
ἐτυμος true (18)
εὐ well (18)
εὐθυς straight (18)
εὑρισκω I discover (20)
εὐρος east wind (7)
εὐρυς broad (18)
εὐωχια feast (11)
ἐχινος hedgehog (22)
ζεφυρος west wind (7)
ζυγον yoke (14)
ζυμη yeast (23B)
ζυμη yeast, leaven (11)
ζωμος broth (11)
ζωνη belt, girdle (12B
ζωον animal (22)
ἡβη youth (17)
ἡδυς sweet (18)
ἡθμος sieve (14)
ἠλεκτρον amber (10)
ἡλιος sun (8)
ἡμερα day (6)
ἡμι- half (2)
ἡνια bridle (14)
ἡπαρ, -ατος liver (21)
ἠπειρος continent (9)
ἡρως teacher (15)
ἠχος sound (17)
ἠως dawn (6)
ἠως morning (6)
θαλαμος bedchamber (13B)
θαλασσα sea (9)
θαλλος shoot (23B)

175

θανατος	death (17)	χθυς	fish (22)
θαυμα	marvel (17)	ἰχνος	footprint (21)
θεατρον	theatre (13A)	καθεδρα	chair (12A)
θελημα	will (17)	καθολικος	general, universal (18)
θεος	god (17)		
θεραπεια	care, attendance (17)	καινος	new (18)
		καθαρσις	purification, cleansing (17)
θερμος	hot (18)		
θερμοτης	heat (5)	κακος	bad (18)
θερος	summer (6)	καλαθος	basket (12A)
θεσις	arrangement (17)	καλαμος	reed (23B)
θεωρια	reflection, contemplation (17)	καλος	beautiful (18)
		καλυμμα	covering (12B)
		καλυπτα	veil (12B)
θηκη	box (12A)	καλυπτω	I cover (20)
θηλη	teat (21)	καμαρα	arch (13B)
θηρ	beast (22)	καμηλος	camel (22)
θιγμα	touch (17)	καμπη	caterpillar (22)
θριξ, τριχος	hair (21)	καμπυλη	curve (3)
θρομβος	clot (21)	καμπυλος	curved (18)
θυμος	thyme (23A)	κανθαρος	beetle (22)
θυρα	door (13B)	κανθος	corner of eye (21)
θυρεος	shield (14)	κανων	rule, rod (16)
θυσανος	tassel (12B)	καπνος	smoke (5)
θωραξ	breastplate (14)	καρδαμον	cress (23A)
ἰατρικος	medical (18)	καρδια	heart (21)
ἰατρος	physician (15)	καρις	shrimp (22)
ἰδιος	individual (18)	καρκινος	crab (22)
ἰδιος	private (18)	καρκινος	crab (21)
ἱερευς	priest (15)	καρπος	fruit (23B)
ἰλυς	mud (5)	καρπος	wrist (21)
ἰοειδης	violet-like (4)	καρυον	nut, nucleus (23B)
ἰος	arrow (14)	καρφος	straw (10)
ἱππος	horse (22)	κασσιτερος	tin (10)
ἰρις	iris (23A)	καστωρ	beaver (22)
ἰρις	rainbow (7)	κατα	down (1)
ἰσθμος	isthmus (9)	κατακλυσμος	flood (9)
ἰσος	equal (18)	καταλυμα	tavern, inn (13A)
ἰστιον	sail (14)	καταρακτης	waterfall (9)
ἱστορια	narrative (17)	καυλος	stalk (23B)
ἰστος	loom (14)	κεδρος	cedar (23A)
ἰστος	mast (14)	κεκτημενος	owner (15)
ἰστος	web (12B)	κενος	empty (18)
ἰσχιον	thigh, hip (21)	κεντρον	sting (21)
ἰσχνος	lean (18)	κεντρον	goad, sting (14)

176

κεντρον	centre (3)	κομψος	elegant (18)
κεραμος	tile (13B)	κονδυλος	knuckle (21)
κεραννυμι	I mix (20)	κονις	dust (10)
κερας	horn (21)	κοπρος	dung (21)
κεραυνος	thunderbolt (7)	κοραξ	crow (22)
κερκις	shuttle (14)	κορις	bug (22)
κερκοπιθηκος	monkey (22)	κορμος	trunk (23B)
κερκος	tail (21)	κορυμβος	cluster of flowers (23B)
κεφαλη	head (21)		
κηπος	garden (13B)	κορυνη	club (14)
κηρος	wax (10)	κορυς	helmet (12B)
κητος	whale (22)	κοσκινον	sieve (14)
κιθαρα	lute (14)	κοσμος	universe, world (8)
κιναρα	artichoke (23A)	κοτυλη	cup (12A)
κινεω	I move (20)	κοφινος	basket (12A)
κινναβαρι	vermilion (10)	κοχλιας	snail (22)
κιστη	box (12A)	κοχλιας	vice (14)
κλαδος	bough, branch (23B)	κραμβη	cabbage (23A)
		κρανιον	skull (21)
κλαω	I break (20)	κρατεω	I govern (20)
κλειθρον	key (14)	κρατηρ	vessel, bowl (12A)
κλεις, -ειδος	key (14)	κρεας	meat, flesh (11)
κλειστος	shut (18)	κρεας	flesh (21)
κλεπτης	thief (15)	κρημνος	cliff (9)
κλιβανος	oven (13B)	κρινον	lily (23B)
κλιμακτηρ	step (13B)	κρινω	I separate (20)
κλιμαξ	staircase (13B)	κρισις	verdict, judgement (15)
κλινη	bed (12A)		
κλινω	I bend (20)	κριτης	judge (15)
κλων	shoot (23B)	κροκοδειλος	crocodile (22)
κνεφας	twilight (6)	κροκος	saffron (23A)
κνημις	boot, legging (12B)	κρυος	cold (17)
		κρυπτω	I hide (20)
κνιδη	nettle (23B)	κρυσταλλος	ice (5)
κογχος	shellfish (22)	κρυσταλλος	crystal (10)
κογχη	shell (21)	κτεις, κτενος	comb (14)
κοιλωμα	pit, cavity (19)	κυανεος	blue (4)
κοιλος	hollow (18)	κυβερνητης	steersman, pilot (15)
κοιμαω	I sleep (20)		
κοινος	common (18)	κυβος	cube (3)
κοκκος	berry, grain (23B)	κυκλος	circle (3)
κοκκυξ	cuckoo (22)	κυκνος	swan (22)
κολεος	sheath (14)	κυλινδρος	cylinder (3)
κολλα	glue (10)	κυμα	wave (5)
κολπος	bay (9)	κυπαρισσος	cypress (23A)

κυριος	master (15)	μαθημα	learning (16)
κυρτος	curved (18)	μακρος	long (18)
κυστις	bladder, bag (21)	μαλακος	soft (18)
κυτος	vessel (12A)	μαλλος	wool (10)
κυων, κυνος	dog (22)	μανια	madness (21)
κωμη	village (15)	μανος	scanty (18)
κωνειον	hemlock (23B)	μαργαριτης	pearl (10)
κωνος	cone (23B)	μαρμαρον	marble (10)
κωπη	oar (14)	μαρσιπος	bag (12A)
λαγως	hare (22)	μαστιγοω	I whip (20)
λαιλαψ	whirlwind (7)	μαστιξ	whip (14)
λαιμος	throat (21)	ματαιος	vain (18)
λαμπας	lamp (12A)	μαχαιρα	knife (12A)
λαμπυρις	glow-worm (22)	μαχη	fight (17)
λαμπω	I shine (20)	μεγας	big (18)
λαρυγξ	gullet (21)	μειων	less (18)
λατρεια	worship (17)	μελας	black (4)
λειος	smooth (18)	μελι	honey (11)
λεκιθος	yolk (21)	μελισσα	bee (22)
λεξις	speech (17)	μερος	part (17)
λεπις	scale (21)	μεσημβρια	noon (6)
λεπτος	thin (18)	μεσος	middle (18)
λευκος	white (4)	μετα	after (1)
λεων	lion (22)	μεταλλεια	mine (5)
λιγνυς	soot (5)	μεταξα	silk (10)
λιθος	stone (10)	μετρεω	I measure (20)
λιμνη	lake, pool (9)	μην	month (6)
λιμος	hunger (11)	μητηρ	mother (15)
λινον	flax, linen (23B)	μηχανη	machine (14)
λινον	flax (10)	μηχανημα	device (14)
λιπος	fat (10)	μικρος	little (18)
λισσος	smooth (18)	μιμησις	imitation (17)
λιτος	simple, single (2)	μινθα	mint (23A)
λογος	discourse (16)	μιξις	mixture (17)
λογος	word (16)	μισθος	reward, prize (15)
λοξος	crosswise (18)	μισος	hatred (17)
λοφος	comb, crest (21)	μιτος	thread (14)
λυκος	wolf (22)	μιτρα	belt, girdle (12B)
λυρα	lyre (14)	μνημη	memory (17)
λυσις	release (17)	μολυβδος	lead (10)
λυχνος	lamp (12A)	μονας	unit (17)
λυχνος	candle (12A)	μονος	alone (18)
μαγνης	magnet (10)	μονος	single (2)
μαγος	magician (15)	μορεα	mulberry (23A)
μαζα	cake (11)	Μορφευς	sleep (17)

μορφη	form, shape (17)	ξενος	stranger, foreigner (15)
μουσειον	study (13B)		
μουσικη	music (17)	ξηρος	dry (18)
μοχλος	lever (14)	ξιφος	sword (14)
μυθος	fable (16)	ξυλον	wood, timber (23B)
μυκης	mushroom (23B)		
μυξα	phlegm, mucus (21)	ξυλον	wood, timber (10)
		ὀγχη	pear (23B)
μυριος	innumerable (18)	ὀδος	road, (9) path (13B)
μυρμηξ	ant (22)	ὀδους, ὀδοντος	tooth (21)
μυς, μυος	mouse (22)	ὀδυνη	pain (17)
μυς, μυος	muscle, mouse (21)	οἰδημα	swelling (21)
		οἰκεω	I dwell (20)
μυσος	dirt (10)	οἰκοδομημα	building (13A)
μυστηριον	secret rite (17)	οἰκονομος	steward (15)
μωρος	foolish (18)	οἰκος	house (13A)
ναος	temple (13A)	οἰνος	wine (11)
ναρκη	numbness (21)	οἰσοφαγος	oesophagus (21)
ναρκισσος	daffodil (23A)	ὀκτω	eight (2)
ναυς	ship, skiff (14)	ὀλιγος	few (18)
ναυσια	sea-sickness (21)	ὀλος	whole (18)
ναυτης	sailor (15)	ὀμβρος	shower (7)
νεκρος	dead (18)	ὀμιχλη	mist (7)
νεος	new (18)	ὀμμα, -ατος	eye (21)
νευρον	nerve, tendon (21)	ὀμος	alike (18)
νεφελη	cloud (8)	ὀμος	similar (18)
νεφος	cloud (8)	ὀνομα	name (17)
νεφρος	kidney (21)	ὀντα	existence (17)
νημα, -ατος	thread (14)	ὀνυξ, ὀνυχος	nail, claw (21)
νησος	island (9)	ὀξος	vinegar (11)
νηχω	I swim (20)	ὀξυς	acid, sharp (18)
νικη	victory (15)	ὀξυς	sharp (18)
νιτρον	saltpetre (10)	ὀπισθε	behind (18)
νομος	custom (15)	ὀπισθεν	backwards (1)
νομος	law (15)	ὀπλον	weapon (14)
νοτος	south (9)	ὀπωρα	autumn (6)
νοτος	south wind (7)	ὀραμα	view, sight (17)
νυκτερις	bat (22)	ὀργανον	tool (14)
νυμφη	bride (15)	ὀργια	secret rite (17)
νυξ, νυκτος	night (6)	ὀρεξις	longing, desire (17)
νωτον	back (21)		
ξανθος	yellow (4)	ὀρθος	straight (18)
ξενος	guest (15)	ὀρισμα	boundary (9)
ξενος	foreign (18)	ὀρμαω	I rouse (20)
ξενος	host (15)	ὀρνις, ὀρνιθος	bird (22)

ὄρος	mountain (9)	πένης	pauper (15)
ὄρχις	orchid (23A)	πέντε	five (2)
ὄρχις	testicle (21)	πεπερι	pepper (23A)
ὀσμη	odour (17)	πεπερι	pepper (11)
ὀστεον	bone (21)	πεπλος	robe (12B)
ὀστρακον	shell (21)	πεπτω	I cook (20)
ὀστρεον	oyster (22)	περδιξ	partridge (22)
ὀσφρα	smell, sense of (21)	περι	around (1)
		περισσος	odd, uneven (18)
οὐδος	threshold (13B)	περιστερα	pigeon (22)
οὐλος	woolly (18)	περονη	pin (14)
οὐρα	tail (21)	πεταλον	leaf, petal (23B)
οὐρανος	heaven (8)	πετρα	rock (5)
οὐρον	urine (21)	πεψις	digestion (21)
οὐς, ὠτος	ear (21)	πηγη	fountain, spring (5)
ὀφθαλμος	eye (21)		
ὀφις	snake (22)	πηδαλιον	rudder (14)
ὀφρυς	eyebrow (21)	πηληξ	helmet (12B)
ὀχημα	vehicle (14)	πηλος	clay (5)
ὀψις	appearance, eyesight (21)	πιθηκος	ape (22)
		πικρος	bitter (18)
ὀψον	food (11)	πιλος	wool (21)
παγη	trap, snare (14)	πιλος	cap (12B)
παγος	frost (7)	πιναξ	dish (12A)
παθος	suffering (17)	πισος	pea (23A)
παις, παιδος	child, boy (15)	πλαγιος	crosswise (18)
παλαιος	old (18)	πλανητης	wanderer (15)
παλη	flour (11)	πλανος	wandering (18)
παλιν	again (18)	πλαξ, πλακος	tombstone (14)
πανοπλια	armour (12B)	πλασις	formation (17)
παπυρος	paper (16)	πλασμα	figure, image (17)
παρα	beside, near (1)		
παρεια	cheek (21)	πλαστος	modelled (18)
παρθενος	virgin (15)	πλατανος	plane tree (23A)
πας, παντος	all (18)	πλατυς	flat (18)
πασσαλος	peg, stake (14)	πλειστος	most (18)
πατεω	I walk (20)	πλειων	more (18)
πατηρ	father (15)	πλευρα	side, rib (21)
παχυς	thick (18)	πλεω	I sail (20)
πεδιον	plain (9)	πληθος	large quantity (17)
πελαγος	sea (9)	πληρης	full (18)
πελαργος	stork (22)	πλησιος	near (18)
πελεκυς	axe (14)	πλινθος	brick (13B)
πελμα	sole (21)	πλουτος	riches (17)
πενης	poor (18)	πνευμα	breath (21)

πνευμα	breath (5)	πυραμις	pyramid (3)
πνευμων	lungs (21)	πυρετος	fever (21)
πνοη	breath (5)	πυριτης	flint (10)
ποα	grass (23A)	πυρρος	reddish (4)
ποιεω	I create (20)	πωγων	beard (21)
ποιημα	poem (16)	πωλεω	I sell (20)
ποικιλος	various (18)	ῥαβδος	stick (23B)
πολεμιοι	enemy (15)	ῥαβδος	rod (14)
πολεμος	war (17)	ῥαφανις	radish (23A)
πολιος	grey (4)	ῥαφη	seam (12B)
πολις	state (15)	ῥαφις	needle (14)
πολις	city (15)	ῥαχις	backbone (21)
πολιτης	citizen (15)	ῥεος	stream (9)
πολυπους	octopus (22)	ῥευμα	current (9)
πολυς	many, much (18)	ῥεω	I flow (20)
πορφυρεος	purple (4)	ῥηγνυμι	I burst (20)
ποσις	drink (11)	ῥητορικη	rhetoric (16)
ποταμος	river (9)	ῥιζα	root (23B)
πους, ποδος	foot (21)	ῥινη	file (14)
πραγμα	fact, deed (17)	ῥιπιζω	I fan (20)
πραξις	action (17)	ῥις, ῥινος	nose (21)
πρεσβυς	old man (15)	ῥοδοεις	pink (4)
πρεσβυς	ambassador (15)	ῥοδον	rose (23B)
πριω, πριζω	I saw (20)	ῥομβος	rhombus (3)
πριων	saw (14)	ῥοπαλον	club (14)
προ	before (1)	ῥυγχος	snout (21)
προβλημα	proposition (17)	ῥυθμος	rhythm (17)
προοιμιον	preface (16)	σακχαρ	sugar (11)
προσκεφαλαιον	pillow, cushion (12A)	σαλαμανδρα	salamander (22)
		σαλαμβη	window (13B)
προφητης	interpreter (15)	σαλπιγξ	trumpet (14)
πρωκτος	anus (21)	σαπρος	putrid (18)
πρωτος	first (2)	σαρξ, σαρκος	flesh (21)
πτερις	fern (23B)	σαυρα	lizard (22)
πτερνα	heel (21)	σεισμος	shaking, shock (17)
πτερον	wing (21)		
πτερυγιον	fin (21)	σελαχος	shark (22)
πτιλον	feather (21)	σεληνη	moon (8)
πυγη	buttocks (21)	σημα	sign, symbol (17)
πυκνος	compact (18)	σημασια	signification (17)
πυλη	gate (13B)	σημειον	sign (17)
πυλωμα	gate (13B)	σηπια	cuttlefish (22)
πυξις	box (12A)	σηπω	I cause to putrefy (20)
πυος	discharge, pus (21)		
πυρ	fire (5)	σιδηρος	iron (10)

σιναπι	mustard (11)	στοα	porch (13A)
σιναπι	mustard (23A)	στολη	robe (12B)
σιτος	food (11)	στομα	mouth (21)
σιφων	siphon (14)	στομαχος	opening of
σκαφη	ship, skiff (14)		stomach (21)
σκελος	leg (21)	στρατηγος	commander (15)
σκια	shadow (8)	στρατιωτης	soldier (15)
σκιουρος	squirrel (22)	στρεπτος	twisted (18)
σκληρος	hard (18)	στροβεω	I spin, whirl (20)
σκομβρος	mackerel (22)	στρογγυλος	round (18)
σκοπεω	I look at (20)	στρουθος	ostrich (22)
σκορπιος	scorpion (22)	στροφευς	hinge (14)
σκοτος	darkness (17)	στροφη	twist (17)
σκυφος	cup (12A)	στρυχνος	nightshade (23B)
σκωληξ	worm (22)	στυπτηρια	alum (10)
σοφια	wisdom (17)	συκον	fig (23A)
σοφος	wise (18)	συλλαβη	syllable (16)
σπαθη	sword, blade (14)	συμ-	together with (1)
σπασμος	spasm (21)	συμβολη	battle (17)
σπερμα	seed (23B)	συμβολη	battle (15)
σπηλαιον	cave (5)	συμμαχος	ally (15)
σπινθηρ	spark (5)	συμπτωμα	symptom (21)
σπλαγχνα	bowels (21)	συν	together with (1)
σπλην	spleen (21)	συριγξ,	whistle (14)
σπογγια	sponge (22)	συριγγος	
σποδος	ashes (5)	σφαιρα	sphere (3)
σπονδυλος	vertebra (21)	σφενδονη	sling, catapult (14)
σπορος	seed (23B)	σφην	wedge (14)
σταγμα	drop (5)	σφυγμος	pulse (21)
σταλαω	I drip (20)	σφυρα	hammer (14)
σταφυλη	bunch (of grapes)	σχιζω	I split (20)
	(23B)	σχισμη	cleft (5)
στεαρ	fat (10)	σχολη	school (16)
στεγη	roof (13B)	σωλην	tube, pipe (14)
στεμμα	crown (12B)	σωμα	body (21)
στενον	strait (9)	ταξις	battle array, rank
στενος	narrow (18)		(15, 17)
στερεος	solid (18)	ταπης	carpet (12A)
στερνον	breast (21)	ταυρος	bull (22)
στεφανος	wreath (12B)	ταφος	grave (13A)
στηθος	chest (21)	ταφρος	ditch (13B)
στηλη	column, pillar	ταχυς	quick (18)
	(13B)	τειχος	wall (13B)
στιγμα	mark (17)	τεκτων	builder (15)
στιχος	row, verse (16)	τελος	end, purpose (17)

182

τεμνω	I cut (20)	τυραννος	dictator (15)
τενων	tendon (21)	τυρος	cheese (11)
τερας	marvel (17)	τυφλος	blind (18)
τερας	omen (17)	ὑακινθος	hyacinth (23A)
τερηδων	timberworm (22)	ὑαλος	glass (10)
τεσσαρες	four (2)	ὑγιεια	health (21)
τετρα-	four (2)	ὑγιης	healthy (18)
τετραγωνον	square (3)	ὑγρος	moist, wet (18)
τεφρα	ashes (5)	ὑδραργυρος	mercury (10)
τεχνη	art (16)	ὑδωρ, ὑδατος	water (5)
τεχνιτης	artisan (15)	ὑετος	rain (5)
τηλε	distant (18)	υἱος	son (15)
τηλε	afar (18)	ὑλη	wood, timber
τιγρις	tiger (22)		(23B)
τιμη	price, cost (15)	ὑλη	forest, wood (9)
τιτανος	lime (10)	ὑλη	wood, timber (10)
τοκος	birth (17)	ὑμην	membrane (21)
τονος	stretching (17)	ὑμνος	hymn (16)
τοξευω	I shoot (arrows)	ὑπερ	beyond (1)
	(20)	ὑπερ	above, over (1)
τοξον	bow (14)	ὑπνος	sleep (17)
τοπος	place (17)	ὑπο	below, under (1)
τραγος	goat (22)	ὑποδημα	slipper, shoe (12B)
τραπεζα	table (12A)	ὑποκριτης	actor (15)
τραυμα	wound (21)	ὑσσωπος	hyssop (23A)
τραχεια	windpipe (21)	ὑστριξ	porcupine (22)
τραχυς	rough (18)	ὑψος	height (17)
τρημα	hole, cavity (17)	φαγειν	I devour (20)
τρια	three (2)	φαινω	I shine (20)
τριβω	I rub (20)	φαινω	I show (20)
τριπλοος	treble (2)	φαιος	grey (4)
τριτος	third (2)	φαλαγξ	joint (of toe or
τρομος	shaking, shock		finger) (21)
	(17)	φαλλος	penis (21)
τροπη	direction, turning	φανερος	visible (18)
	(17)	φαντασμα	apparition (15)
τροπις	keel (14)	φαρετρα	quiver (14)
τροφη	food (11)	φαρμακον	drug (21)
τροχιλια	pulley (14)	φαρυγξ	throat (21)
τροχος	wheel (14)	φασιανος	pheasant (22)
τρυμα	hole, cavity (17)	φασις	speech (17)
τρυπανον	gimlet (14)	φιλεω	I love (20)
τυλος	lump, callus (17)	φλεγω	I burn (20)
τυμπανον	drum (14)	φλεψ, φλεβος	vein (21)
τυπος	model, type (17)	φλοξ, φλογος	flame (5)

φοβεομαι	I fear (20)	χιτων	tunic (12B)
φοβεω	I frighten (20)	χιων	snow (7)
φορεω	I carry (20)	χλαμυς	cloak (12B)
φορος	tax, tribute (17)	χλωρος	green (4)
φραγμα	hedge, fence (13B)	χοανη	funnel (14)
φρασις	phrase (17)	χολη	bile (21)
φρην	understanding (16)	χονδρος	cartilage (21)
φρυνη	toad (22)	χοριον	skin, leather (21)
φυκος	seaweed (23B)	χοριον	leather (10)
φυλακτηρ	guard (15)	χορος	dance (16)
φυλλον	leaf (23B)	χορτος	hay (10)
φυλον	tribe, clan (15)	χρεος	debt (15)
φυσα	bellows (14)	χρημα	thing, object (17)
φυσαλις	bubble (5)	χρηματα	money, wealth
φυσις	nature (17)		(15)
φυτον	plant (23B)	χρονος	time (6)
φυω	I grow (20)	χρυσος	gold (10)
φωκαινα	porpoise (22)	χρωμα	colour (4)
φωκη	seal (22)	χρως, -ωτος	skin (21)
φωνη	sound (17)	χυλος	juice (11)
φως, φωτος	light (17)	χυμος	juice (11)
χαιτη	mane, long hair	χωριον	ground, soil (5)
	(21)	χωρος	space (17)
χαλαζα	tubercle (21)	ψαλμος	song (17)
χαλαζα	hail (7)	ψαμμος	sand (5)
χαλκος	copper (10)	ψευδω	I deceive (20)
χαμαι	on the ground (5)	ψηφος	pebble (5)
χασμα	cleft (5)	ψιττακη	parrot (22)
χειλος	lip (21)	ψυλλα	flea (22)
χειμων	storm (7)	ψυχη	butterfly (22)
χειμων	winter (6)	ψυχη	mind (17)
χειρ	hand (21)	ψυχρος	cold (18)
χειροκτιον	glove (12B)	ωδη	song (17)
χελωνη	tortoise (22)	ωδινω	I suffer (20)
χερσονησος	peninsula (9)	ωκεανος	ocean (9)
χηλη	talon (21)	ωον	egg (21)
χην	goose (22)	ωρα	hour (6)
χθων	ground, soil (5)	ωσμος	thrust (17)
χιλιοι	thousand (2)	ωψ	eye (21)